T0332398

A Student's Guide to the Ising Model

The Ising model provides a detailed mathematical description of ferromagnetism and is widely used in statistical physics and condensed matter physics. In this Student's Guide, the author demystifies the mathematical framework of the Ising model and provides students with a clear understanding of both its physical significance and how to apply it successfully in their calculations. Key topics related to the Ising model are covered, including exact solutions of both finite and infinite systems, series expansions about high and low temperatures, mean-field approximation methods, and renormalization-group calculations. The book also incorporates plots, figures, and tables to highlight the significance of the results. Designed as a supplementary resource for undergraduate and graduate students, each chapter includes a selection of exercises intended to reinforce and extend important concepts, and solutions are also available for all exercises.

James S. Walker received his Ph. D. in theoretical physics at the University of Washington. He is Emeritus Professor of Physics at Washington State University, and has held positions at the University of Pennsylvania and the Massachusetts Institute of Technology. He has taught the Ising model at both the undergraduate and graduate level and is the author of the highly successful undergraduate text *Physics* (Addison-Wesley) now in its fifth edition.

Other Books in the Student's Guide Series:

A Student's Guide to the Ising Model

JAMES S. WALKER
Washington State University

CAMBRIDGE
UNIVERSITY PRESS

Shaftesbury Road, Cambridge CB2 8EA, United Kingdom

One Liberty Plaza, 20th Floor, New York, NY 10006, USA

477 Williamstown Road, Port Melbourne, VIC 3207, Australia

314–321, 3rd Floor, Plot 3, Splendor Forum, Jasola District Centre,
New Delhi – 110025, India

103 Penang Road, #05–06/07, Visioncrest Commercial, Singapore 238467

Cambridge University Press is part of Cambridge University Press & Assessment,
a department of the University of Cambridge.

We share the University's mission to contribute to society through the pursuit of
education, learning and research at the highest international levels of excellence.

www.cambridge.org
Information on this title: www.cambridge.org/9781009098519

DOI: 10.1017/9781009089579

First published 2023

A catalogue record for this publication is available from the British Library.

ISBN 978-1-009-09851-9 Hardback
ISBN 978-1-009-09630-0 Paperback

Additional resources for this publication at www.cambridge.org/walker-sgim.

This book owes its existence to the Three Amigos—
Betsy Walker, Janet Walker, and Jennifer Knudson. I love you!

Acknowledgements

I would like to thank the editorial and production staff at Cambridge University Press—Nicholas Gibbons, Sarah Armstrong, Elle Ferns, Jane Chan, and Reshma Xavier—for making the entire process of producing this book a most enjoyable experience. And to the students who will learn about the Ising model with this book, I hope you will come to love and appreciate it as much as I have.

About This Book

This edition of *A Student's Guide to the Ising Model* is supported by solutions to all problems and Mathematica files, available via the book's website. Please visit www.cambridge.org/walker-sgim to access this extra content.

We may update our Site from time to time and may change or remove the content at any time. We do not guarantee that our Site, or any part of it, will always be available or be uninterrupted or error free. Access to our Site is permitted on a temporary and "as is" basis. We may suspend or change all or any part of our Site without notice. We will not be liable to you if for any reason our Site or the content is unavailable at any time, or for any period.

Contents

1

The Ising Model

Few models in theoretical physics have been studied for as long, or in as much detail, as the Ising model. It's the simplest model to display a nontrivial phase transition, and as such it plays a unique role in theoretical physics. In addition, the Ising model can be applied to a wide range of physical systems, from magnets and binary liquid mixtures, to adsorbed monolayers and superfluids, to name just a few. In this chapter, we present some of the background material that sets the stage for a detailed study of the Ising model in the chapters to come.

1.1 Magnetic Phase Transitions

The original purpose of the Ising model was to study phase transitions in magnetic systems. We'll start, then, with a brief discussion of magnetic phases and phase diagrams, and we'll introduce the Ising model itself in the next section.

A simple magnetic system, like an iron magnet for example, displays two distinct behaviors, as illustrated in Figure 1.1 (a). At high temperatures the system is *paramagnetic*. In this region, the magnetization of the iron "tracks" with the applied magnetic field, B. What this means is that if the magnetic field is positive, relative to a particular direction, the magnetization is positive as well; if the magnetic field is zero, the magnetization is zero; if the magnetic field is negative, the magnetization is negative. In fact, for small magnetic fields the magnetization is simply proportional to the field.

At low temperatures the system is *ferromagnetic*. As the magnetic field is reduced from positive to zero in this region, the magnetization decreases to a finite positive amount and remains at that value. This is a magnetization that is due to the interactions between individual electron spins, and it persists even when there is no external field to align the spins. Similarly, as the field is

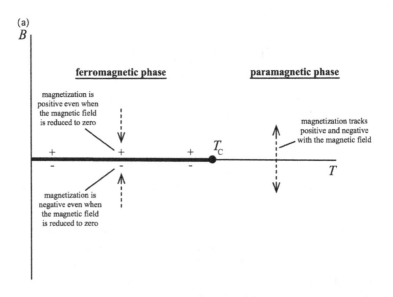

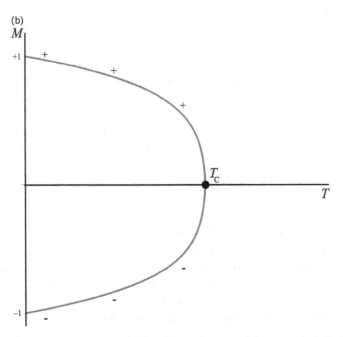

Figure 1.1 (a) A magnetic phase diagram in terms of the magnetic field, B, and temperature, T. At high temperature the system is paramagnetic, with no permanent magnetization. At low temperature the system is ferromagnetic, with up (+) and down (−) magnetization coexisting along the $B = 0$ line. The critical point at temperature T_c separates the ferromagnetic and paramagnetic regions. (b) Zero-field magnetization, M, versus temperature, T, for the system in part (a). Notice that the discontinuity in magnetization vanishes at the critical point, T_c.

reduced in magnitude from the negative side, the magnetization approaches a finite negative value. These observations are indicated by the + and − signs above and below the temperature axis in Figure 1.1 (a).

It follows that there are two coexisting phases of matter at zero magnetic field – the positive and negative magnetization phases – and that a discontinuity in the magnetization occurs as the magnetic field crosses $B = 0$. No such discontinuity is observed in the paramagnetic region of the phase diagram. The discontinuity in the magnetization decreases in magnitude with increasing temperature, and eventually vanishes at the critical temperature, T_c.

To highlight the key role played by the magnetization, consider a plot of magnetization, M, versus temperature, T, as in Figure 1.1 (b). In this plot, the applied magnetic field is zero, $B = 0$. As a result, the magnetization of the system is zero at temperatures that are higher than T_c. When the temperature is reduced to a value just below T_c, the system becomes magnetized; we refer to this as a *spontaneous magnetization*. Whether the spontaneous magnetization is positive or negative is determined randomly by the majority of spins when the temperature drops below T_c. If we reduce the temperature further, the result is an increasing magnitude of the magnetization. Eventually the magnetization saturates to its largest positive or negative value, which we normalize as $M = \pm 1$. The "jump" from the lower to the upper branch of the magnetization curve is the discontinuity that characterizes ferromagnetic behavior.

The way to think about this phase transition on a microscopic level is to picture electron spins in the iron atoms trying to align with their neighbors through exchange interactions. At high temperatures, thermal fluctuations overwhelm the magnetic interactions and the spins are oriented randomly. As the temperature is lowered, the magnetic interactions begin to win out over the thermal fluctuations, and the system spontaneously picks an orientation for the entire system. We say that the initial up–down symmetry of the system (recall that $B = 0$) has been "spontaneously broken" as a finite up or down magnetization takes over.

The magnetization is referred to as the *order parameter* for this phase transition. In general, the order parameter of a system is a quantity that is zero in the disordered phase, and finite for temperatures below T_c. In addition, the order parameter is a measure of the *amount* of order in the system; the larger the order parameter, the greater the order.

A mathematical function with a discontinuity, like the one we see in the magnetization below T_c, is referred to as a nonanalytic, or singular function – a key feature of phase transitions. It follows that the system's free energy – the function that contains all of the equilibrium information about the system – must be singular as well, as are its various derivatives. Since the magnetization is the first derivative of the free energy with respect to the magnetic field, phase transitions like this are sometimes referred to in older literature as "first-order" phase transitions. The accepted modern terminology is to simply refer to the jump in magnetization below T_c as a "discontinuous" phase transition.

In contrast, consider the phase transition that occurs in Figure 1.1 (b) as one moves from high to low temperature along the temperature axis. In this case, the magnetization rises smoothly from zero to nonzero values as the temperature goes through T_c. This sort of phase transition, with a continuous change in the value of the order parameter, is referred to as a "continuous" phase transition.

Another example of a continuous phase transition is shown in Figure 1.2. Here we see the temperature dependence of the zero-field magnetic suscepti-bility, χ, which is defined as the derivative of the magnetization, M, with respect to the magnetic field, B:

$$\chi = \frac{\partial M}{\partial B}.$$

We see that χ is a smooth function, with no discontinuities. It is still a singular function, however; in fact, notice that it diverges to infinity at the critical temperature, T_c.

Again, we see that a phase transition is characterized by singular behavior in a thermodynamic function. The susceptibility χ is the derivative of M with respect to B, and M is the derivative of the free energy with respect to B, so χ is

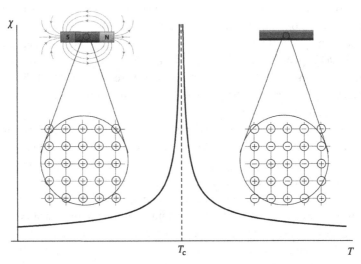

Figure 1.2 The zero-field magnetic susceptibility, $\chi = \partial M/\partial B$, as a function of temperature. The susceptibility blows up to infinity at the critical point, T_c.

the second derivative of the free energy. As a result, older literature sometimes refers to this as a "second-order" phase transition.

We shall return frequently to the topic of the free energy, which we denote by f, in our study of the Ising model. In fact, we will calculate the free energy for many different systems, and we'll see that it is the key to understanding the behavior of a system. Once we obtain the free energy, we take first derivatives to find average values, and second derivatives to find fluctuations about the average. This will be a recurring theme throughout our exploration of the Ising model.

Comparing with Other Phase Diagrams

Now that we've looked carefully at the characteristics of a magnetic phase diagram, we'd like to put it in context by comparing with the phase diagram of a familiar three-phase substance. In Figure 1.3 (a) we show a typical phase diagram for a material with solid, liquid, and gaseous phases. The bold curves denote discontinuous phase transitions, where two phases coexist along a curve in the pressure–temperature plane. As these curves are crossed, there is a discontinuity in the density of the material. The exception is at the critical

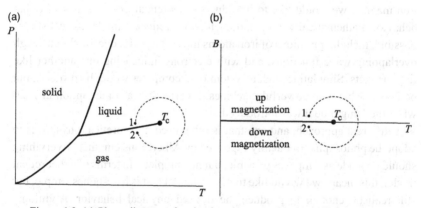

Figure 1.3 (a) Phase diagram of a simple substance with three distinct phases: solid, liquid, and gas. The bold curves denote the coexistence of the two phases on either side of the curve. The liquid–gas coexistence curve terminates at the critical point T_c, beyond which there is no distinction between liquid and gas. Points 1 and 2 can be connected without a phase transition by following a curved path, like the dashed circle, that goes around the critical point. The solid–liquid curve extends to infinity. (b) The magnetic phase diagram is similar to the liquid–gas portion of the phase diagram in part (a), with up magnetization and down magnetization playing the roles of liquid and gas, respectively. Points 1 and 2 can be connected without a phase transition.

point, T_c, where the density difference between the liquid and gas phases vanishes – similar to the vanishing magnetization difference at T_c in the magnetic phase diagram in Figure 1.3 (b).

In fact, the behavior along the liquid–gas transition curve in Figure 1.3 (a) and the up-magnetization/down-magnetization line in Figure 1.3 (b) are analogous in many respects. In both cases, a discontinuity is encountered if the bold line is crossed. On the other hand, the system can be brought smoothly, and with no discontinuities at all, along the dashed paths connecting points 1 and 2. The critical points, labeled T_c, are points where a continuous phase transition is observed.

The similarity in behavior near these critical points is more than just qualitative – on close examination, the quantitative details associated with the singularities are precisely the same in both the magnetic and nonmagnetic substances. These surprisingly deep connections, which lead to the concept of *universality* in critical behavior, will be explored in detail in Chapter 5.

1.2 The Ising Model of Magnetism

Now that we're familiar with the magnetic behavior of a simple system like an iron magnet, we would like to introduce a theoretical model to describe that behavior mathematically. One approach is to produce a model as realistic as possible, including millions of iron atoms interacting with one another through overlapping wave functions, and with electrons influencing one another like tiny magnets. Simulating such a system on a computer would be difficult, but perhaps doable with powerful machines. It certainly wasn't an option in 1920 when the Ising model was introduced.

A different approach, and one that is often used in theoretical physics, is to adopt the philosophy, frequently espoused by Albert Einstein, that "Everything should be made as simple as possible, but not simpler." In terms of a theoretical model, this means we would like to have a model that is bare-bones simple, but still realistic enough to produce the desired physical behavior. Additional complications can be added later, if desired, to include other types of behavior; but it's best to study the most basic aspects of a system first. The Ising model is a good example of this approach.

Ising Variables

The Ising model is constructed with "spin" variables, s, that occupy the sites, i, of a lattice. The value of the variable at any given lattice site, s_i, is meant to represent, in a crude way, the spin of an electron. Now, as we know, electrons are spin-1/2 particles, with magnetic moments that are either aligned or anti-aligned along a specified direction in space. In the Ising model, we represent this by saying that each spin is either up (+1) or down (–1). Thus, Ising spins are simply dimensionless numbers:

$$s_i = \pm 1.$$

More than anything else, a two-valued variable like this is what identifies a model as "Ising like." The variables can interact in various ways, or lie on the vertices of various lattices, but they always have this simple plus–minus/up–down/yes–no quality.

A microstate (or state, or configuration) of an Ising model consists of a specific assignment of +1 or –1 to each of the spins in the system. If the system is a lattice with N sites, it follows that the total number of states is 2^N. In the coming chapters, we'll carry out sums over all such states to obtain the thermodynamic functions associated with the Ising model. A specific example of spin assignments for a group of Ising spins is given in Figure 1.4, along with different ways of representing the spins pictorially.

The variables in an Ising model can be given different names, as long as they have the same basic Ising symmetry. For example, consider a binary-liquid mixture, consisting of two types of molecules, A and B. We can describe a system like this with an Ising model whose variables represent one type of molecule, $s_i = A$, or the other, $s_i = B$, as in Figure 1.5 (a). Similarly, the Ising model could represent a system of helium atoms adsorbed on a surface of graphite. In this case, each hexagonal adsorption site on the graphite surface is either occupied by a helium atom ($s_i = 1$) or empty ($s_i = 0$), as illustrated in Figure 1.5 (b).

These are just a couple examples of the many nonmagnetic systems that have an underlying Ising-like character to them, and can be studied with the Ising model. We will generally speak of the Ising model in magnetic terms, but it should be remembered that the comments can apply to other types of systems as well.

Ising Hamiltonians

The next step in constructing an Ising model is to introduce interactions involving the spin variables. These interactions produce different energies for different spin states, as described by the Hamiltonian of the system. For example, the Hamiltonian might have an energy term related to a spin aligning or antialigning with an external magnetic field. Similarly, the Hamiltonian could include a term describing the interaction of one spin with its neighbors.

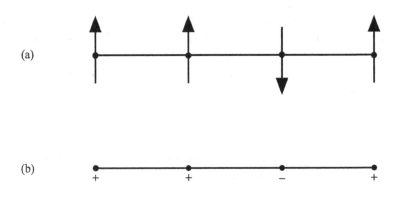

Figure 1.4 (a) Ising spins are often represented as up or down arrows indicating spins that are +1 or −1, respectively. In the state shown here, the spin values, reading from left to right, are +1, +1, −1, +1. (b) Ising spins can be represented with + or minus − signs. (c) A different way of representing the same state as in parts (a) and (b).

To begin, consider a single spin, s_1, interacting with an external magnetic field, B. The Hamiltonian H in this case can be written as follows:

$$H = -\mu_B B s_1.$$

In this expression, μ_B is the Bohr magneton, which means that $\mu_B B$ has the units of energy. The minus sign in front of μ_B ensures that a positive magnetic field, $B > 0$, favors a positive spin, $s_1 = +1$. To see why this is so, consider the energy levels of the two states of the spin:

Spin state	Energy
$s_1 = +1$	$H = -\mu_B B$
$s_1 = -1$	$H = \mu_B B$

Notice that the lowest energy for $B > 0$ corresponds to $s_1 = +1$, and so this state is favored. If $B < 0$ the lowest energy occurs when $s_1 = -1$. Thus, with this choice for the Hamiltonian, a given sign of B favors the same sign of s_1.

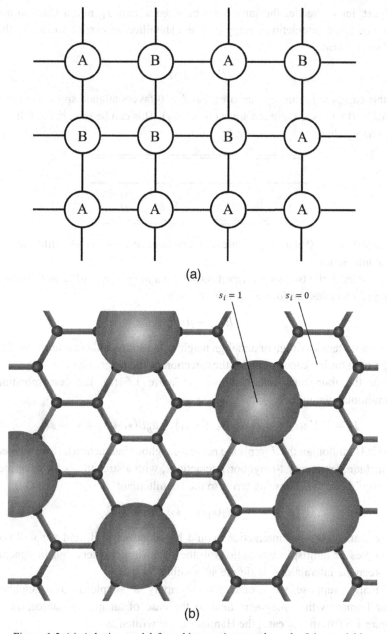

(a)

$s_i = 1$ $s_i = 0$

(b)

Figure 1.5 (a) A lattice model for a binary mixture, where the Ising variables take on the values $s_i = A$ or $s_i = B$. (b) The centers of the hexagons on a graphite lattice are adsorption sites that can be occupied by a helium atom, $s_i = 1$, or empty, $s_i = 0$. We can think of the s_i as Ising variables.

Next, let's consider the interaction between a spin, s_1, and a neighboring spin, s_2. If the interaction energy is J, the Hamiltonian can be written in the following form:

$$H = -Js_1s_2.$$

In this case, the minus sign ensures that $J > 0$ favors aligned spins ($s_1 = s_2$), and $J < 0$ favors antialigned spins ($s_1 = -s_2$). This can be seen in the following energy chart:

Spin state	Energy
$s_1 = s_2$	$H = -J$
$s_1 = -s_2$	$H = J$

We say that $J > 0$ is a ferromagnetic interaction, and $J < 0$ is an antiferromagnetic interaction.

Combining the two-spin interaction J with a magnetic field term B for two Ising spins yields the following Hamiltonian:

$$H = -Js_1s_2 - \mu_B B(s_1 + s_2).$$

The J term tends to align or antialign neighboring spins, and the B terms tend to align or antialign each spin with the direction of the magnetic field.

For the four-spin system shown in Figure 1.6 (a), the corresponding Hamiltonian would be

$$H = -J(s_1s_2 + s_2s_3 + s_3s_4 + s_4s_1) - \mu_B B(s_1 + s_2 + s_3 + s_4).$$

In this Hamiltonian the J terms are nearest-neighbor interactions. If we wanted to include next-nearest-neighbor interactions, with a strength J_2 for example, we would add the following terms to the Hamiltonian:

$$-J_2(s_1s_3 + s_2s_4).$$

Three- and four-spin interactions could be added as well, and we will see examples of multispin interactions in the following chapters, but in general we keep the interactions as simple as possible.

Finally, suppose we would like to apply a simple nearest-neighbor Hamiltonian with a magnetic field to the case of an infinite lattice, as in Figure 1.6 (b). In this case, the Hamiltonian is written as

$$H = -J \sum_{\langle ij \rangle} s_i s_j - \mu_B B \sum_i s_i. \tag{1.1}$$

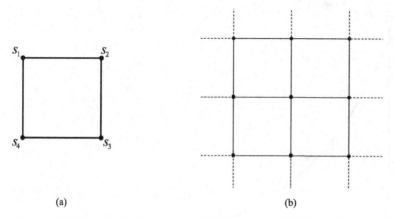

Figure 1.6 (a) A four-spin Ising system. (b) A section of an infinite square lattice.

The first term indicates a sum over nearest-neighbor pairs (denoted $\langle ij \rangle$) on the lattice. The second term is a straightforward sum over all sites i of the lattice.

Now that we have a Hamiltonian that gives an energy for every state of the system – that is, for every assignment of $+1$ and -1 to the spins – how do we use this information to study the Ising model? We address this question in the next two sections.

1.3 The Boltzmann Factor

The key to statistical mechanics, and to the Ising model in particular, is the Boltzmann factor. To understand the role of the Boltzmann factor, imagine a system in thermal equilibrium at a fixed absolute temperature T – this is referred to as the canonical ensemble. The system undergoes thermal fluctuations from state to state. The probability that a given state n occurs is proportional to the exponential of $-E_n/k_B T$, where E_n is the energy of the state and k_B is Boltzmann's constant, $k_B = 1.38 \times 10^{-23}$ joule/kelvin. Stated mathematically, the probability can be written as follows:

$$\text{probability} \propto e^{-E_n/k_B T}. \tag{1.2}$$

This exponential, which is plotted in Figure 1.7, is referred to as the Boltzmann factor.

We will convert the expression for probability to an equality shortly, but for now note that the probability decreases with increasing energy for a fixed

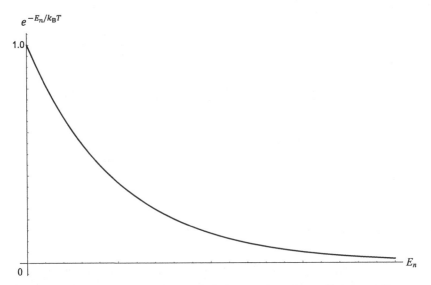

Figure 1.7 At constant temperature, the Boltzmann factor drops off exponentially
with increasing energy, E_n.

temperature. In fact, it is the ratio of the energy, E_n, to the *thermal energy*, $k_B T$,
that determines the probability of a state. You can think of it this way: thermal
equilibrium at the temperature T corresponds to the thermal energy $k_B T$. This
amount of energy is available for excitations in the system. If the energy of
a given state is many times greater than the thermal energy, there is little chance
that random thermal fluctuations will produce the state – in fact, an exponentially decreasing chance.

Boltzmann Factors and Probabilities in a Two-State System

To get a better feeling for the Boltzmann factor, consider a simple system with
just two states, one with energy E_1 and the other with energy E_2. The probability P_1 that the state with energy E_1 occurs is

$$P_1 = Ae^{-E_1/k_B T}.$$

Notice that we've introduced a normalizing factor, A, to produce an equality.
Similarly, the probability for state 2 is

$$P_2 = Ae^{-E_2/k_B T}.$$

Now, the sum of all the probabilities must equal 1, so we have

$$P_1 + P_2 = Ae^{-E_1/k_B T} + Ae^{-E_2/k_B T} = A(e^{-E_1/k_B T} + e^{-E_2/k_B T}) = 1.$$

It follows that the normalizing factor is

$$A = 1/(e^{-E_1/k_B T} + e^{-E_2/k_B T}).$$

We can now write explicit expressions for the probabilities:

$$P_1 = \frac{e^{-E_1/k_B T}}{e^{-E_1/k_B T} + e^{-E_2/k_B T}}$$

$$P_2 = \frac{e^{-E_2/k_B T}}{e^{-E_1/k_B T} + e^{-E_2/k_B T}}.$$

These probabilities are plotted in Figure 1.8 for the case $E_2 > E_1$.

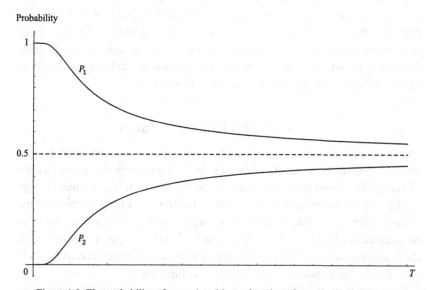

Figure 1.8 The probability of states 1 and 2 as a function of temperature, T. As the temperature goes to zero, the lowest-energy state – state 1 in this case – is occupied with unit probability, and there is no chance of finding the system in the higher-energy state, 2. As the temperature goes to infinity, the difference in energy between the two states is insignificant, and hence the states are equally likely to be occupied; thus, the probability is ½ for both states in this limit.

Introducing the Partition Function

Notice that the denominators of P_1 and P_2 are simply the sum of the Boltzmann factors for all states of the system. This sum plays such an important role in statistical mechanics that we give it a special name, the *partition function*, Z. Thus, in this case,

$$Z = e^{-E_1/k_B T} + e^{-E_2/k_B T}.$$

Generalizing to a system with an arbitrary number of states, labeled with n and having energies E_n, we can write

$$Z = \sum_{\text{all states, } n} e^{-E_n/k_B T}. \tag{1.3}$$

The corresponding probabilities are

$$P_n = \frac{e^{-E_n/k_B T}}{Z}. \tag{1.4}$$

In one sense, the partition function can be thought of as simply a normalization factor, giving normalized probabilities. There's more to it than that, however. As we shall see, the partition function, with its sum of Boltzmann factors over all states, contains *all* of the equilibrium thermodynamic information about a system. Not surprisingly, calculating Z will be a prime objective of our calculations going forward.

Motivating the Boltzmann Factor

The Boltzmann factor is crucial in all the calculations that follow, so let's take a moment to consider some of its main attributes. First, the probability of a state of energy E_n should depend on the ratio of the energy to the thermal energy; $E_n/k_B T$. Next, the larger the value of this ratio, the lower the probability of the corresponding state. There are many mathematical functions that satisfy these requirements, like $(E_n/k_B T)^{-1}$ or $(E_n/k_B T)^{-2}$, for example. However, of all the functions one could write that decrease with an increasing ratio of $E_n/k_B T$, there is one that stands out – the exponential function.

Consider, then, the following expression for the probability of a state with energy E_n:

$$P_n = \frac{1}{Z} e^{-E_n/k_B T}.$$

We know this is correct, but for the moment let's view it as a possibility that we'd like to investigate to see if it has the desired properties for a probability distribution. It certainly depends on the ratio $E_n/k_B T$, and it decreases with an increasing value of the ratio. So far, so good.

Now, imagine a system comprised of two independent subsystems, 1 and 2. Subsystem 1 has two energies, E_{11} and E_{12}, and subsystem 2 has two energies, E_{21} and E_{22}. The partition function for the composite system is a sum over the four possible states:

$$Z = e^{-(E_{11}+E_{21})/k_B T} + e^{-(E_{11}+E_{22})/k_B T} + e^{-(E_{12}+E_{21})/k_B T} + e^{-(E_{12}+E_{22})/k_B T}.$$

Because of the nature of exponential functions, each term can be rewritten as the product of two separate terms; for example, $e^{-(E_{11}+E_{21})/k_B T} = e^{-E_{11}/k_B T}e^{-E_{21}/k_B T}$. It follows that Z can be factored into a product of two partition functions, one for each subsystem:

$$Z = (e^{-E_{11}/k_B T} + e^{-E_{12}/k_B T})(e^{-E_{21}/k_B T} + e^{-E_{22}/k_B T}) = Z_1 Z_2.$$

The fact that the partition function of a system with independent subsystems is equal to the product of the partition functions for each subsystem will be useful in chapters to come.

We can now answer questions like the following: what is the probability that subsystem 1 is in state 2 and subsystem 2 is in state 1? The total energy of these states is $E_{12} + E_{21}$, and hence the desired probability is

$$P = \frac{1}{Z}e^{-(E_{12}+E_{21})/k_B T} = \frac{1}{Z_1 Z_2}e^{-E_{12}/k_B T}e^{-E_{21}/k_B T} = P_{12}P_{21}.$$

Thus, an *exponential* probability distribution gives the desired result that the probability of independent states occurring is equal to the product of the probabilities for each state occurring separately. This provides strong support for the supposition that the Boltzmann factor should be an exponential function. We would not have obtained this result with a different functional dependence on the energy.

Next, consider a physical system that we move from the basement to the third floor of a building. In so doing, all the states acquire an additional energy E_0 due to the increased gravitational potential energy at the new location. This energy changes the Boltzmann factor for each state – and it also changes the partition function. In Problem 1.7 we show that the net result, due to the exponential dependence on energy, is that the probabilities for the states are unchanged. It follows that the physics of the system is unchanged by this relocation – as one would expect. This, again, is strong support for an exponential probability distribution.

Now that we've motivated the exponential form of the Boltzmann factor, one might wonder why we write the exponential as $e^{-E_n/k_B T}$ instead of using some other base, like 2. Well, in fact, we could use 2, or any other base. For example, you could write the Boltzmann factor as $2^{-E_n/kT}$, where k has a different numerical value than k_B. The value of k for base 2 is determined in Problem 1.8. While adjusting the value of k would allow us to use 2 as the base, it would result in additional numerical factors when taking derivatives and doing other mathematical operations. Using the base e allows us to take advantage of its convenient mathematical properties, but it isn't strictly necessary.

Thermal Energy

Earlier in this section, when we introduced the Boltzmann factor, we referred to $k_B T$ as the thermal energy. The thermal energy determines the range of energies that are important for a given range of temperatures. Let's evaluate the thermal energy for an everyday situation.

To be specific, consider room temperature. Recalling that T is the absolute temperature, we set $T \sim 300$ K. Substituting this into $k_B T$ yields an energy of roughly 1/40 eV. Thus, the molecules in the air in your room are moving with typical energies of 0.025 eV. Recalling that the difference in energy levels in an atom is on the order of eV, we see that the Boltzmann factor for atomic excitation is roughly $e^{-40} \sim 10^{-18}$. Thus, we can be assured that all the molecules in the air are in their lowest energy state – there simply isn't enough thermal energy available to excite them.

Converting the thermal energy to kinetic energy, it follows that the typical speeds of oxygen and nitrogen molecules in the atmosphere are in the range of 500 m/s – comparable to the speed of sound, of course. This is much lower than the escape velocity of the Earth ($v = 11,200$ m/s), so the Earth retains its oxygen and nitrogen. If we calculate the thermal speed of hydrogen molecules, however, which have much less mass and therefore greater speed for the same energy, we find $v \sim 1,700$ m/s. This speed is still well short of the escape speed, but it's higher than the speeds for oxygen and nitrogen and is large enough to give a reasonable probability of escaping the Earth. Because of thermal energy, then, there is little hydrogen left in Earth's atmosphere today – what hydrogen the Earth possessed initially has been lost in space.

1.4 Solving the Ising Model

In much of physics, "solving" a problem means obtaining the solution to a differential equation. Prime examples are solutions to Newton's laws of motion in mechanics, Maxwell's equations in electromagnetism, and Schrödinger's equation in quantum mechanics. When we study the Ising model, however, there are no differential equations. Evidently, "solving" the Ising model is something quite different from what we're used to in other branches of physics.

So, how do we solve the Ising model? In short, the Ising model is solved when we obtain the partition function, Z. Once we know Z, we use it to calculate thermodynamic quantities like the free energy, the average energy, and the specific heat. Let's take a look at how this is done.

The basic connection between statistical mechanics, which deals with probabilities, and thermodynamics, which deals with energy transformations, is contained in the following expression for entropy:

$$S = -k_{\mathrm{B}} \sum_{\text{all states}, n} P_n \ln P_n. \tag{1.5}$$

In this expression, the quantities P_n are the probabilities, written in terms of the Boltzmann factor and the partition function:

$$P_n = \frac{e^{-E_n/k_{\mathrm{B}}T}}{Z}.$$

A completely equivalent way of writing Equation (1.5), and one that is more convenient for our purposes going forward, is the following:

$$F = E - TS = -k_{\mathrm{B}}T \ln Z. \tag{1.6}$$

In this expression, F is the Helmholtz free energy, E (without subscripts) is the average energy, T is the absolute temperature, and S is the entropy. Notice that F is related directly to the logarithm of the partition function.

The reason we focus on the free energy is that, in many ways, it is analogous to the *wave function* ψ in quantum mechanics. In quantum systems, once you know the wave function, you can use it to extract all the information that can be known about a system. For example, taking a derivative of the wave function gives an average value, like $\langle x \rangle$; taking two derivatives gives fluctuations, or uncertainties, like Δx.

Similarly, the free energy, F, which is also referred to as the *fundamental relation*, plays an analogous role in statistical mechanics. Specifically, the free energy contains all the equilibrium information that can be obtained about the system. As in quantum mechanics, taking a derivative of F yields an average value, like E for example, and taking two derivatives yields a fluctuation, like the specific heat, C. The name free energy, which we will use as a shorthand for the fundamental relation, has historical roots that refer to energy that is "free," or "available," to do useful work.

To gain some insight into the expression $F = -k_B T \ln Z$ and how it relates to $F = E - TS$, let's take a look at some special cases. First, consider a simple system that has only a single state with the energy E_1. It follows that the partition function for this system is

$$Z = e^{-E_1/k_B T}.$$

Calculating the free energy, we find

$$F = E - TS = -k_B T \ln Z = -k_B T(-E_1/k_B T) = E_1.$$

This is what we expect for this system – that is, the average energy E is simply E_1, and there is no entropy since there is just one state. The reason for the factor $-k_B T$ preceding the natural log of the partition function is now clear.

Next, suppose our system has Ω states, all with the energy E_1. Now the partition function is

$$Z = \Omega e^{-E_1/k_B T}.$$

The corresponding free energy is

$$F = E - TS = -k_B T \ln Z = -k_B T(\ln \Omega - E_1/k_B T) = E_1 - k_B T \ln \Omega.$$

The entropy of a system with Ω states of fixed energy is

$$S = k_B \ln \Omega. \tag{1.7}$$

In addition, the average energy of this system is $E = E_1$. It follows that $F = E - TS = E_1 - k_B T \ln \Omega$, as expected.

One final point about $F = E - TS$ is that it represents a competition between energy and entropy, since these quantities enter F with opposite signs. In general, a system in equilibrium tends to minimize its free energy. At low temperatures the entropy term is negligible, and minimizing F is the same as minimizing the energy, E. At high temperatures, on the other hand, the temperature term dominates, and minimizing F means maximizing the entropy, S. These opposite extremes are embodied within F.

Reduced Quantities and Notation

The inverse of the thermal energy occurs so often that we give it a compact, special name, β:

$$\beta = \frac{1}{k_B T}.$$

The Boltzmann factor, then, can be written as $e^{-\beta E_n}$. Note that β is inversely proportional to the temperature.

Clearly, it would be convenient to include a factor of $-\beta$ in our definition of the Hamiltonian since it comes up so often in the Boltzmann factors. Recall that a simple Hamiltonian for the Ising model, containing both a nearest-neighbor coupling and a magnetic field, is

$$H = -J \sum_{\langle ij \rangle} s_i s_j - \mu_B B \sum_i s_i.$$

Each of the terms in this equation has the dimensions of energy. In theoretical calculations, it is generally preferable to use dimensionless, or *reduced*, quantities. We refer to $-\beta H$ as the *reduced Hamiltonian* and write it as follows:

$$-\beta H = \left(\frac{J}{k_B T}\right) \sum_{\langle ij \rangle} s_i s_j + \left(\frac{\mu_B B}{k_B T}\right) \sum_i s_i.$$

To simplify this expression, we define a dimensionless nearest-neighbor coupling as $K = J/k_B T$ and a dimensionless magnetic field as $h = \mu_B B/k_B T$. With these definitions we have

$$-\beta H = K \sum_{\langle ij \rangle} s_i s_j + h \sum_i s_i. \tag{1.8}$$

This form of the Hamiltonian appears many times in the coming chapters.

If we imagine the coupling K to result from an interaction energy J that is characteristic for a given system, then K varies inversely with temperature. In the limit $T \to \infty$, the thermal energy overwhelms the interaction energy, and the coupling in Equation (1.8) goes to zero; $K \to 0$. When $T \to 0$, the interaction energy dominates the system, and $K \to \infty$.

The expression in Equation (1.8) is in a form that's ready to be used in the Boltzmann factor. For example, suppose we have a system with just two neighboring spins, s_1 and s_2. In this case, the reduced Hamiltonian is

$$-\beta H = K s_1 s_2 + h(s_1 + s_2). \tag{1.9}$$

If $s_1 = +1$ and $s_2 = +1$, the corresponding Boltzmann factor is

$$e^{-\beta H} = e^{K+2h}.$$

Reduced quantities like these are common in studies of the Ising model.

The free energy, F, can be cast in a dimensionless form as well. Recall from Equation (1.6) that

$$F = -k_B T \ln Z.$$

Multiplying by $-\beta$ makes this a dimensionless quantity. In addition, note that the free energy is an extensive function, meaning that it scales with the size of the system – in our case, it scales with the number of spins, N. We would like to work with a function that measures the free energy *per spin*, so that our results don't depend on the number of sites. Thus, we define a reduced free energy per site as follows:

$$f = -\beta F / N = \frac{1}{N} \ln Z. \tag{1.10}$$

It is f that we use most often in our theoretical calculations.

One final bit of notation: Note that we use i to label a site on a lattice, and s_i to label the spin s on the site i. To denote a *state* of the system – which is just an assignment of $+1$ or -1 for each spin s_i – we will use a special notation that is common in the field of statistical mechanics. This notation is $\{s_i\}$, and it simply means a state of the system:

$$\{s_i\} = \text{a state of the system.}$$

A sum over all states is written as follows:

$$\sum_{\{s_i\}} = \text{a sum over } all \text{ states.}$$

Using this notation, the partition function of a system is

$$Z = \sum_{\{s_i\}} e^{-\beta H(\{s_i\})}.$$

In this expression we write $H(\{s_i\})$ to show explicitly that the Hamiltonian depends on the state of the system. When this dependence is clear from the context, and doesn't need to be emphasized, we'll just write H to simplify the notation.

As an explicit example of how the sum over states works, consider a system with just two spins, and with the interactions described by the reduced Hamiltonian in Equation (1.9). The sum over all states for this system is

$$\sum_{\{s_i\}} e^{-\beta H} = \sum_{\{s_i\}} e^{Ks_1s_2+h(s_1+s_2)} = \sum_{s_1\in\{+1,-1\}} \sum_{s_2\in\{+1,-1\}} e^{Ks_1s_2+h(s_1+s_2)}.$$

The symbols $s_1 \in \{+1,-1\}$ and $s_2 \in \{+1,-1\}$ indicate that these two sums are over the elements of the set $\{+1,-1\}$; that is, the first of these sums is over $s_1 = +1$ and $s_1 = -1$, and the second is over $s_2 = +1$ and $s_2 = -1$. The result of these summations is as follows:

$$\sum_{s_1\in\{+1,-1\}} \sum_{s_2\in\{+1,-1\}} e^{Ks_1s_2+h(s_1+s_2)} = e^{K+2h} + e^{-K} + e^{-K} + e^{K-2h}$$
$$+ + \qquad + - \qquad - + \qquad - - .$$

The spin configuration for each term in the double summation has been indicated with plus and minus signs below the corresponding term. For example, we use the symbols $+ -$ to indicate the configuration where $s_1 = +1$ and $s_2 = -1$. The corresponding Boltzmann factor is e^{-K}.

It's interesting to note that summing over all states yields a result – the partition function – that does not depend on the state, but rather includes information from *all* the states. The situation is similar to integrating a function of x and y with respect to x; the result is a function that depends only on y – the x dependence has been integrated out. In our case, the states have been integrated (summed) out, leaving a result that no longer depends on the state of the system.

Obtaining a Solution to the Ising Model

To wrap things up for this section, we can summarize the procedure to solve the Ising model as follows:

Steps to Solving the Ising Model
1. Choose a group of Ising spins to study. This can be a finite number of spins, or an infinite number on a lattice in one or more dimensions.
2. Choose a reduced Hamiltonian $-\beta H$ that produces a dimensionless energy for every spin configuration.
3. Sum the Boltzmann factors over all the spin configurations $\{s_i\}$ to obtain the partition function, Z:

$$Z = \sum_{\{s_i\}} e^{-\beta H}.$$

4. Calculate the reduced free energy per site f from the partition function:

$$f = \frac{1}{N} \ln Z.$$

5. Use f to calculate thermodynamic quantities of interest.

We'll put these steps into action numerous times in the following chapters.

1.5 A Brief History of the Ising Model

The history of the Ising model is interesting in its own right, with several twists and turns over the years. As mentioned earlier, it was invented in 1920 by Wilhelm Lenz (1888–1957) as a model for magnets and their phase transitions. Lenz assigned the problem of solving the model to his graduate student Ernst Ising (1900–1998) for his Ph.D. project. Ising published his results for the infinite, one-dimensional lattice, with nearest-neighbor couplings and a magnetic field, in 1925 – in the only research paper he published. Known ever since as the Ising model (pronounced today as "eye zing" rather than the more correct "ee zing"), it has developed into many forms since its introduction. In fact, Ising's single paper is surely one of the most referenced scientific papers in recent times.

Good News/Bad News

The good news for Ising was that he was able to solve the problem – no mean feat in the 1920s. Recall that each spin has two possible states (+1 and –1), so a model with N spins has a total of 2^N states. For an infinite lattice, the number of states goes to infinity exponentially, so a solution requires some clever mathematics. We cover Ising's exact solution in Chapter 4, and also show some simpler, more modern approaches.

The bad news was that the model failed in its main goal, which was to produce a phase diagram like that in Figure 1.1. Ising found that the model does not have a phase transition at finite temperature: to be generous, one might say that the critical point in Figure 1.1 is suppressed to zero temperature in the 1-D Ising model. In any case, the solution to the model was surely a great disappointment.

Ising was faced with several possible explanations for why the model failed to show a finite-temperature phase transition. A few of the more likely ones are as follows: (1) It might be that the variables in the model are too simple to represent a real magnet. (2) Perhaps 1-D is too low a dimension to show a phase transition. (3) Maybe longer-range interactions are needed. Ising took the view that the model was too simple, and/or the interactions were too short range. He expressed the view that the model would not show a phase transition in any dimension because of these deficiencies. It turns out that, happily for the model, Ising was incorrect.

The next bit of good news for the Ising model was provided by Rudolf Peierls (1907–1995) in 1936. Peirels was able to show that the Ising model in two and higher dimensions would indeed show a finite transition temperature, and hence a phase diagram like that in Figure 1.1. It turns out that the simple variables and nearest-neighbor only interactions are still quite compatible with phase transitions – the real culprit in the original study was the low dimension. We'll explore the effect of dimensionality in Chapter 5. Peierls's result breathed new life into the Ising model because now it was certain that additional work in studying higher-dimensional models would be worth the effort.

The Ising model became even more interesting in 1941 when Hans Kramers (1894–1952) and Gregory Wannier (1911–1983) showed that high- and low-temperature series expansions are related in a way that yields the precise location of the phase transition in 2-D. This is referred to as *duality*, a concept we'll also explore in Chapter 5. Using duality, Kramers and Wannier were able to show that the phase transition in the 2-D nearest-neighbor Ising model on a square lattice occurs when the coupling K has the following critical value:

$$K_c = \frac{1}{2}\ln(1 + \sqrt{2}) = 0.4406\ldots.$$

This numerical value is one that is familiar to all who study the Ising model.

So now, thanks to Kramers and Wannier, it was known not only that a finite-temperature phase transition occurs in the 2-D model, but also the exact location of the transition. There was still no solution for the full model, but that was soon to come.

Exact Solution

Our understanding of the Ising model took a quantum leap forward when Lars Onsager (1903–1976) published the exact solution of the zero-field, nearest-neighbor Ising model on a two-dimensional square lattice in 1944. Onsager's solution, like Ising's, involved the use of transfer matrices, which are discussed in detail in Chapter 4. The solutions differ greatly, however, because the transfer matrix in Ising's case is 2×2, whereas in Onsager's case it is infinity by infinity. To complete the solution, Onsager needed to find the largest eigenvalue of this infinite-by-infinite matrix, which he did in what is justly referred to as a mathematical tour de force.

Onsager received the Nobel Prize in Chemistry in 1968. The award is officially for his work on reciprocal relations involved in diffusion, but clearly it also recognizes his achievements with the Ising model. The situation is similar to that of Albert Einstein winning the Nobel Prize for the photoelectric effect, though his work on relativity is much better known.

It should be stressed that Onsager's solution is confined to zero field ($h = 0$); even so, it contains the most important behavior of the model. For example, it has a critical point at $K_c = 0.4406...$, in agreement with the prediction of Kramers and Wannier. At that location there are singularities in the thermo-dynamic functions. Perhaps the most dramatic of these is the logarithmic divergence to infinity of the specific heat. There have been many attempts to extend the Onsager solution to finite field, but so far none have succeeded.

According to stories that must surely be apocryphal, Onsager assigned his graduate student the task of extending his two-dimensional solution to the case of the three-dimensional Ising model. If there is any truth to the story, one can only hope the graduate student quickly changed to a different thesis topic, otherwise he or she would still be at it – no one has yet solved the 3-D model. The only successful extensions of the Onsager solution are to a few closely related 2-D models, like the zero-field Ising model on a triangular or hexagonal lattice.

Thus, the Onsager solution, as brilliant as it is, is basically a one-time shot. It solves the problem but is not the basis for a wide variety of solutions to other systems. As a result, while Onsager's solution is important and of great interest, it is not a route to further results for the Ising model.

The vast majority of Ising systems studied today are explored with one or more types of approximation techniques. Many of these approximation methods, like the mean-field approximations and the renormalization-group techniques, are of considerable interest in their own right, and are best under-stood when studying their application to the Ising model. Thus, the Ising model is an ideal "test bed" for exploring various types of approximation techniques.

Universality

The Ising model is "simple" in the sense that its underlying variables have just two values, and the interactions among these variables are limited. Even so, the two-dimensional model, which is difficult to solve despite its apparent simpli-city, exhibits a nontrivial finite-temperature phase transition that can be studied with full mathematical rigor. For this reason alone, it deserves a prominent place in the annals of theoretical physics.

This only scratches the surface, however, when it comes to reasons why physicists are intrigued by the Ising model. As already mentioned, the Ising model can be used to study an impressive variety of systems with an up–down/left–right type of symmetry, like liquid–vapor critical points, mixing–unmixing transitions in binary-liquid mixtures, coil–uncoil transitions in DNA, and the normal–super transition in superfluids. Thus, the Ising model is of interest partly because of its wide applicability.

An even deeper reason for the Ising model's continued relevance is the property of universality, mentioned earlier. Simply put, universality refers to the fact that different systems, with very different physics, have the same behavior in certain key respects near their critical points. For example, the way thermodynamic functions like the specific heat diverge to infinity has universal properties that show an unexpected connection between ferromagnets and liquid mixtures.

As we understand it today, universality reflects the fact that a system's behavior near a critical point is determined by the general symmetry of its interacting variables, and by the dimensionality of the system – not by the details of the interactions. The fact that critical behavior exhibits universality is one of the main reasons physicists are so interested in the subject today – it indicates that something very fundamental is going on in these systems.

Thus, the Ising model is of even greater interest than one might have thought. It applies not only to magnetic systems, but to a wide variety of different physical systems. In fact, it defines an entire *universality class* in critical phenomena. Lenz and Ising could never have imagined the incredible insight that would flow from their simple model, but they would surely be pleased to see the intensive research that it continues to inspire.

1.6 Problems

1.1 Figure 1.9 shows a B-versus-T plot with various numbered points indicated for consideration. (a) At which of the points 1, 2, 3, and 4 is the

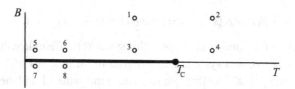

Figure 1.9 A B-versus-T plot with various numbered points from 1 to 8.

magnetization the greatest? (b) Is the magnetization at point 2 greater than, less than, or equal to the magnetization at point 4? (c) Is the magnetization at point 3 greater than, less than, or equal to the magnetization at point 4?

1.2 Figure 1.9 shows a B-versus-T plot with various numbered points indicated for consideration. (a) At which of the points 5, 6, 7, and 8 is the magnetization the greatest? (b) Is the difference in magnetization in going from point 6 to point 5 greater than, less than, or equal to the difference in magnetization in going from point 7 to point 5?

1.3 Two nearest-neighbor Ising systems have identical coupling energies, J, but system 1 has a reduced coupling K_1 and system 2 has a reduced coupling K_2. If $K_1 > K_2$, which system has the higher temperature?

1.4 The critical point of the nearest-neighbor, two-dimensional square lattice is given by

$$K_c = 0.4406 \ldots$$

Do you expect K_c for the nearest-neighbor, two-dimensional triangular lattice to be greater than, less than, or equal to $0.4406 \ldots$? Explain.

1.5 Two Ising spins are connected by a nearest-neighbor bond, as indicated in Figure 1.10. The reduced Hamiltonian is

$$-\beta H = Ks_1 s_2 + h(s_1 + s_2).$$

(a) What is the number of states in this system? (b) Calculate the partition function, Z, for this system. Express your result in terms of K and h. (c) What is the probability that both spins have the value $+1$?

$$s_1 \qquad\qquad\qquad\qquad\qquad\qquad s_2$$

Figure 1.10 A group of two Ising spins. The reduced Hamiltonian indicates the spins have a nearest-neighbor coupling, K, and experience a magnetic field, h.

1.6 A group of three Ising spins is arranged in an equilateral triangle, as shown in Figure 1.11. The reduced Hamiltonian is

$$-\beta H = K(s_1 s_2 + s_2 s_3 + s_3 s_1) + h(s_1 + s_2 + s_3).$$

(a) What is the number of states in this system? (b) Calculate the partition function, Z, for this system. Express your result in terms of K and h. (c) What is the probability that two spins have the value $+1$ and the third spin has the value -1?

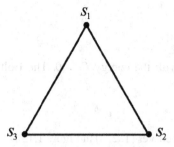

Figure 1.11 A group of three Ising spins. The reduced Hamiltonian indicates the spins have a nearest-neighbor coupling, K, and experience a magnetic field, h.

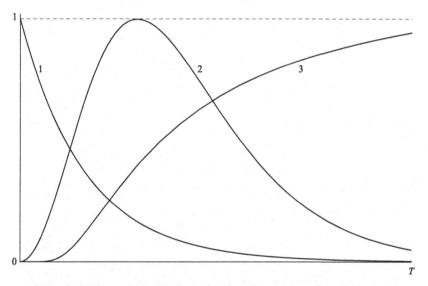

Figure 1.12 Three curves (1, 2, 3) as a function of temperature. One of the curves is the Boltzmann factor.

1.7 Suppose an energy E_0 is added to the energy of each state of a system. (a) In this new system, show that the Boltzmann factor for a state with an original energy E_i can be written as $e^{-E_0/k_B T} e^{-E_i/k_B T}$. (b) Show that the partition function for the new system can be written as $e^{-E_0/k_B T} Z$, where Z is the original partition function. (c) Show that the probability for any given state is unchanged by the added energy.

1.8 Suppose we would like to write the Boltzmann factor in base 2, as follows:

$$2^{-E_i/kT}.$$

Find the value of k.

1.9 Consider a state with the energy $E_i > 0$. The Boltzmann factor for this state is

$$e^{-E_i/k_B T}.$$

Which of the three curves (1, 2, 3) in Figure 1.12 is a plot of the Boltzmann factor as a function of temperature, T?

2

Finite Ising Systems

In this chapter, we explore Ising systems that consist of just one or a few spins. We define a Hamiltonian for each system, and then carry out straightforward summations over all the spin states to obtain the partition function. No phase transitions occur in these systems – in fact, as we show later in the chapter, an infinite system is needed to produce the singularities that characterize phase transitions. Even so, our study of finite systems will yield several results and insights that are important to the study of infinite systems.

2.1 A Single Spin

We begin with the simplest possible Ising system – a single spin, which we denote by s_1. The system is illustrated in Figure 2.1.

There are no other spins for s_1 to interact with, but it can be acted on by an external magnetic field, h. It will then have a reduced (dimensionless) energy, hs_1, and this energy will depend on whether the spin is aligned or antialigned with the magnetic field. It follows that the reduced Hamiltonian is

$$-\beta H = hs_1.$$

This is a simplified, one-spin version of the reduced Hamiltonian given in Equation 1.8.

Calculating the Partition Function and Free Energy

The first step in studying an Ising system is to calculate the partition function, Z, by summing over Boltzmann factors. Specifically, we want to calculate the following:

$$\bullet\, s_1$$

Figure 2.1 A single-spin Ising system, with spin s_1.

$$Z = \sum_{\{s_1\}} e^{-\beta H}. \tag{2.1}$$

In this expression, the symbol $\{s_1\}$ indicates that the sum is over the two values (i.e., *configurations*) of s_1. When we carry out the sum over the values $s_1 = +1$ and $s_1 = -1$ we find

$$Z = \sum_{\{s_1\}} e^{hs_1} = e^h + e^{-h} = 2\cosh h. \tag{2.2}$$

This is the partition function in terms of the interaction strength h.

Notice that we've written the partition function in two different forms. In one, we've expressed Z in terms of a sum of exponentials, which come directly from the Boltzmann factors. In the other, we've written Z in terms of hyperbolic trigonometric functions; in this case, in terms of the hyperbolic cosine, $\cosh h = (e^h + e^{-h})/2$. Hyperbolic functions and their derivatives come up frequently in Ising model calculations. We will use either exponentials or hyperbolic functions in the expressions we write, depending on which is most convenient in a given situation.

Now that we've obtained the partition function, the next step is to calculate the reduced free energy per site, f. Recall from Equation 1.10 that

$$f = \frac{1}{N}\ln Z. \tag{2.3}$$

With $N = 1$ in this case, and Z from Equation 2.2, we have

$$f = \ln(e^h + e^{-h}) = \ln(2\cosh h) = \ln 2 + \ln(\cosh h). \tag{2.4}$$

This result is plotted with the solid curve in Figure 2.2. The sloped dashed line is a plot of $y = h$ for comparison. We see that f is a smooth curve, with no striking features. This is generally the case for free energies, even in systems like the two-dimensional nearest-neighbor model, where the free energy is a singular function. It's usually only after taking two derivatives of the free energy, as in calculating the specific heat, that the singularity becomes apparent.

Checking Limits

We can gain some insight into the free energy by considering various limits. First, let $h \to 0$, where $h = \mu_B B / k_B T$. This limit can be seen in two different

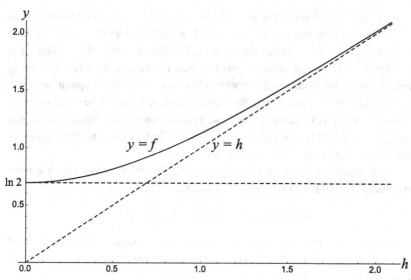

Figure 2.2 The reduced free energy f (solid curve) versus the reduced magnetic field, h, for a single-spin system. At $h = 0$ the free energy is equal to $\ln 2 = 0.693\ldots$ (horizontal dashed line); in the limit $h \to \infty$ the free energy approaches $y = h$ (sloping dashed line).

ways: (1) it is the limit of zero magnetic field, $B = 0$, and hence zero energy, at a finite temperature, T; or (2) it is the limit of infinite temperature, $T \to \infty$, with a finite magnetic field, B. In either case, the magnetic energy is negligible compared with the thermal energy.

Letting $h \to 0$ in Equation (2.4), we find

$$f \to \ln(1 + 1) = \ln 2 = 0.693\ldots.$$

This is a significant result – it is the reduced entropy per site of a system in which each site has two equally likely states. Recall from thermodynamics that

$$f = (-\beta/N)F = (-\beta/N)(E - TS) = (-\beta E/N) + (S/k_B N). \qquad (2.5)$$

In this case, the energy is zero, $E = 0$, and the number of sites is one, $N = 1$. In addition, the number of states is two, and hence the entropy is given by the Boltzmann constant times the log of the number of states:

$$S = k_B \ln 2.$$

Substituting these results into Equation (2.5), we see that the reduced free energy per site is $f = \ln 2$. Thus, we can say that the free energy of the system is all entropy and no energy when $h \to 0$.

Next, let's check the opposite limit, $h \to \infty$. Again, this can be thought of in two ways: (1) an infinite magnetic field at finite temperature; or (2) zero temperature with a finite magnetic field. Either way, the system has a magnetic energy that overwhelms the thermal energy. As a result, there is just one state that occurs in the system – the state in which the spin aligns with the magnetic field – and hence the reduced entropy is zero. The corresponding reduced energy is h. Substituting these results into the thermodynamic equation, $f = (-\beta E/N) + (S/k_B N)$, yields $f = h$ – in this case the free energy of the system is all energy and no entropy.

For comparison, let $h \to \infty$ in Equation (2.4), which comes directly from statistical mechanics and the partition function. This gives

$$f = \ln(e^h + e^{-h}) \to \ln(e^h + 0) = \ln e^h = h.$$

Again, we see complete agreement between the thermodynamic interpretation of the free energy in Equation (2.5) and the statistical mechanical interpretation in Equation (2.3).

We summarize the results for these two limits here:

$h \to 0$ $T \to \infty$	$f \to \ln 2 = 0.693\ldots$	all entropy, no energy
$h \to \infty$ $T \to 0$	$f \to h$	all energy, no entropy

Both of these limits are shown in Figure 2.2.

Magnetization

Let's continue our investigation of this system with a look at the average value of the spin, which we denote by $\langle s_1 \rangle$. We define this to be the magnetization of the system, m:

$$m = \text{magnetization} = \langle s_1 \rangle.$$

We've already learned a bit about m from the limits just considered. For example, in the limit $h \to 0$ the system is random, and hence the spin is equally likely to be $+1$ and -1. As a result, the average value of the spin is zero; $m = \langle s_1 \rangle = 0$. On the other hand, in the limit $h \to \infty$ the system is in the state where the spin is aligned with the magnetic field, and hence $m = \langle s_1 \rangle = 1$ for the case where h is positive, and $m = \langle s_1 \rangle = -1$ when h is negative. Let's extend these results to all values of h.

To do so, we refer to Equation (1.4), which gives the probability that a given state occurs. This result is

$$P(\{s_i\}) = \frac{1}{Z}e^{-\beta H(\{s_i\})}.$$

From this, it follows that if a quantity Q depends on the state of the system, its average value $<Q>$ is simply the sum of the value of Q in each state, times the probability the state occurs. That is,

$$\langle Q \rangle = \sum_{\{s_i\}} Q(\{s_i\})P(\{s_i\}) = \frac{1}{Z}\sum_{\{s_i\}} Q(\{s_i\})e^{-\beta H(\{s_i\})}. \qquad (2.6)$$

Notice that the summation in this expression differs from the partition function only in that the factor $Q(\{s_i\})$ multiplies the Boltzmann factor.

Now, in this case, we would like to set Q equal to the spin, s_1. This gives

$$m = \langle s_1 \rangle = \frac{1}{Z}\sum_{\{s_i\}} s_1 e^{hs_1}. \qquad (2.7)$$

To evaluate m, we carry out the summation and divide by the partition function given in Equation (2.2), $Z = e^h + e^{-h}$. The summation is as follows:

$$\sum_{\{s_i\}} s_1 e^{hs_1} = (+1)e^h + (-1)e^{-h} = e^h - e^{-h}.$$

Combining these results yields the magnetization:

$$m = \frac{e^h - e^{-h}}{e^h + e^{-h}} = \tanh h. \qquad (2.8)$$

Here we have introduced the hyperbolic tangent, tanh, to simplify the notation. This result is plotted in Figure 2.3.

It's clear that the limits previously mentioned agree with Equation (2.8). For example, when $h = 0$, the magnetization is zero as well. In fact, the magnetization "tracks" with the field near $h = 0$; that is, m is proportional to h for small fields, as one would expect for a paramagnetic system. In the limits $h \to +\infty$ and $h \to -\infty$ the magnetization saturates at its extreme values, $m = +1$ and $m = -1$, respectively.

A Derivative of the Free Energy

Take a close look at the expression for the average value of the spin:

$$\langle s_1 \rangle = \frac{1}{Z}\sum_{\{s_1\}} s_1 e^{hs_1}. \qquad (2.9)$$

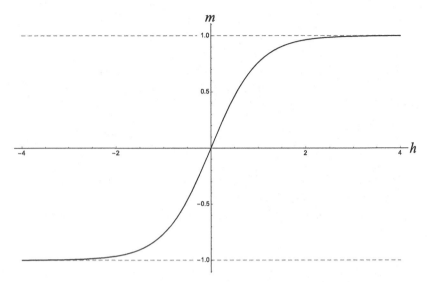

Figure 2.3 The magnetization, $m = \langle s_1 \rangle$, as a function of the reduced magnetic field, h. For $h \to +\infty$ and $h \to -\infty$ the magnetization approaches $+1$ and -1, respectively.

First, notice that s_1 multiplies e^{hs_1} in the summation. A derivative of e^{hs_1} with respect to h would yield just such a term. That is,

$$\frac{\partial e^{hs_1}}{\partial h} = s_1 e^{hs_1}.$$

This is one of the special properties of the exponential function – a derivative with respect to one of its variables brings down the *conjugate variable* (the variable it multiplies) and leaves the exponential unchanged. In this case, we took a derivative with respect to h and brought down its conjugate variable, s_1.

With this result in mind, we see that the summation in Equation (2.9) is actually the derivative of the partition function with respect to h. That is,

$$\frac{\partial Z}{\partial h} = \frac{\partial}{\partial h} \left(\sum_{\{s_1\}} e^{hs_1} \right) = \sum_{\{s_1\}} s_1 e^{hs_1}.$$

It follows that we can write the average of s_1 as

$$\langle s_1 \rangle = \frac{1}{Z} \frac{\partial Z}{\partial h}.$$

We're not done yet. Recall that the derivative of the logarithm of a function is the inverse of the function times the derivative of the function. Therefore, the average value of the spin can be written in the following way:

$$\langle s_1 \rangle = \frac{\partial}{\partial h} \ln Z = \frac{1}{Z} \frac{\partial Z}{\partial h}.$$

This is nothing more than the derivative of the free energy f with respect to the magnetic field:

$$\langle s_1 \rangle = \frac{\partial f}{\partial h}. \tag{2.10}$$

This result shows that the statistical expression for the magnetization, as a sum over probabilities, is equivalent to the thermodynamic expression, which states that the magnetization per site is the rate of change of the reduced free energy per site with respect to the magnetic field:

$$m = \frac{\partial f}{\partial h}. \tag{2.11}$$

Let's quickly verify that our new definition of the magnetization does indeed agree with our previous result. First, recall that the free energy is

$$f = \ln(e^h + e^{-h}).$$

Taking the derivative with respect to h yields

$$\frac{\partial f}{\partial h} = \frac{\partial}{\partial h} \ln(e^h + e^{-h}) = \frac{1}{e^h + e^{-h}} \frac{\partial}{\partial h}(e^h + e^{-h}) = \frac{e^h - e^{-h}}{e^h + e^{-h}} = \tanh h.$$

It's interesting that we can arrive at exactly the same result from two very different types of calculations and two very different perspectives.

In some cases, we'll find it convenient to calculate the average value of a quantity with a sum over probabilities, as in Equation (2.7). In other cases, it will be easier to take the derivative of the free energy with respect to the conjugate variable of the quantity of interest, as in Equation (2.10). We'll use one, the other, or both methods on occasion, depending on the circumstances.

Independent Spins

In Chapter 1, we showed that one of the important properties of the exponential Boltzmann factor is that the partition function for independent systems is just the product of the partition functions for each system separately. For example, consider the system of N spins shown in Figure 2.4.

$$\bullet\,{}^{s}\!1 \qquad \bullet\,{}^{s}\!2 \qquad \bullet\,{}^{s}\!3 \qquad \cdot \qquad \cdot \qquad \cdot \qquad \bullet\,{}^{s}\!N$$

Figure 2.4 N independent Ising spins.

Each spin interacts with an external magnetic field, h, but not with any other spin. As a result, the partition function for each spin is just what we obtained in Equation (2.2):

$$Z = e^{h} + e^{-h}.$$

It follows that the partition function for all N spins, Z_{total}, is

$$Z_{total} = Z^{N}.$$

It's easy to check this explicitly for small values of N. In fact, it's instructive to write out the results for $N = 2$ and $N = 3$ by hand, just to verify the details. These results quickly make it apparent that this expression is valid for arbitrary values of N.

Now that we have the partition function, Z_{total}, let's calculate the reduced free energy per site for the system of N independent spins. We find the following:

$$f = \frac{1}{N}\ln Z_{total} = \frac{1}{N}\ln Z^{N} = \ln Z = \ln(e^{h} + e^{-h}).$$

Thus, the free energy *per site* is the same for N independent spins as it is for a single spin.

We see that having more spins in the system doesn't make it more likely to have a phase transition, at least not if the spins don't interact with one another. It's the interactions that are necessary for the cooperative behavior at the core of phase transitions.

2.2 Two Spins

We turn now to the simplest case of interacting spins; namely, two spins, s_1 and s_2, with a nearest-neighbor interaction, K. The system is illustrated in Figure 2.5.

The reduced Hamiltonian for this system is

$$-\beta H = K s_1 s_2.$$

Recall that $K > 0$ corresponds to a ferromagnetic interaction that tends to make s_1 and s_2 have the same sign, and $K < 0$ corresponds to an antiferromagnetic interaction that tends to make s_1 and s_2 have opposite signs.

Figure 2.5 Two Ising spins with a nearest-neighbor coupling K.

Calculating the Partition Function and Free Energy

The partition function for this system is the sum of Boltzmann factors over all four spin states. The result is

$$Z = \sum_{\{s_i\}} e^{-\beta H} = \sum_{\{s_i\}} e^{Ks_1 s_2} = e^K + e^{-K} + e^{-K} + e^K = 2(e^K + e^{-K})$$

$$\underbrace{++\qquad +-\qquad -+\qquad --}_{\text{Spin configurations}}.$$

The spin configurations for each term in the partition function have been indicated with plus and minus signs below the corresponding term. Thus, for example, $+-$ indicates the configuration $s_1 = +1$ and $s_2 = -1$, and the corresponding Boltzmann factor is e^{-K}.

Next, we calculate the reduced free energy per site, using this result for Z and noting that $N = 2$ for a two-spin system. We find

$$f = \frac{1}{N}\ln Z = \frac{1}{2}\ln 2(e^K + e^{-K}) = \frac{1}{2}\ln(4\cosh K). \qquad (2.12)$$

A plot of f is given in Figure 2.6. As usual, f is a rather simple-looking function, but the limits as $K \to 0$ and $K \to \infty$ have interesting physical interpretations.

First, let's take a look at $K \to 0$, which corresponds to $T \to \infty$. This limit yields

$$K \to 0 \qquad f \to \ln 2 \qquad \text{all entropy, no energy}$$
$$T \to \infty.$$

As with the single-spin system, the infinite-temperature limit corresponds to a completely random system, with no energy, and a reduced entropy per site equal to $\ln 2$; that is, each site has two states. It follows that the intercept of f with the vertical axis is $\ln 2$.

Next, let $K \to \infty$, which correspond to the ground state as $T \to 0$:

$$K \to \infty \qquad f \to \frac{1}{2}\ln 2 + \frac{1}{2}K \qquad \text{ground-state entropy and energy}$$
$$T \to 0.$$

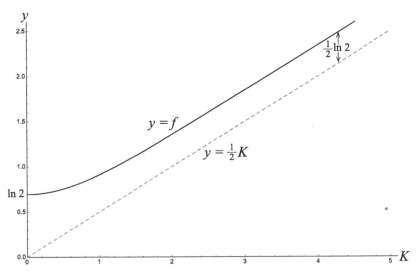

Figure 2.6 The solid curve gives the reduced free energy per site, f, for a two-spin system with coupling K. The dashed line is $y = K/2$. The limiting value of the vertical separation between f and the dashed line is $\frac{1}{2}\ln 2$.

Things are a bit different in this case. First, the spins are aligned since the coupling is infinitely strong. This means that the reduced energy *per site* is K divided by 2.

Second, we notice a new entropy term equal to one-half $\ln 2$. The origin of this term becomes clear when we note that there are two ground states at $T = 0$ — one is when both spins are $+1$, the other is when both spins are -1. Thus, there is a ground-state entropy per site of $\frac{1}{2}\ln 2$. This accounts for the vertical displacement between f and the dashed line ($K/2$) in Figure 2.6.

The Spin–Spin Correlation Function and Average Energy

Now that we have a system with an interaction between spins, it's of interest to see how this interaction acts to correlate the value of one spin to the other. We can explore this *spin–spin correlation* by calculating the average value of the product of the spins, $\langle s_1 s_2 \rangle$. Referring to Equation (2.6), which defines the average value of any quantity, it follows that the average of $Q = s_1 s_2$ is

$$\langle s_1 s_2 \rangle = \frac{1}{Z}\sum_{\{s_i\}} s_1 s_2 e^{-\beta H} = \frac{1}{Z}\sum_{\{s_i\}} s_1 s_2 e^{K s_1 s_2}. \tag{2.13}$$

The summation is over the four states of the spins. Carrying out the sum, and dividing by Z, we find

$$\langle s_1 s_2 \rangle = \frac{1}{Z} \sum_{\{s_i\}} s_1 s_2 e^{K s_1 s_2} = \frac{e^K - e^{-K} - e^{-K} + e^K}{e^K + e^{-K} + e^{-K} + e^K} = \frac{2(e^K - e^{-K})}{2(e^K + e^{-K})} = \tanh K.$$

This result is plotted in Figure 2.7. Notice that $\langle s_1 s_2 \rangle \to +1$ in the ferromagnetic limit $K \to +\infty$ and that $\langle s_1 s_2 \rangle \to -1$ in the antiferromagnetic limit $K \to -\infty$. There is no correlation between the spins when $K \to 0$, and hence $\langle s_1 s_2 \rangle \to 0$ in that case. Notice that these results are the same as those for $\langle s_1 \rangle$ versus h in Figure 2.3.

Let's look at the $K \to 0$ limit a bit more carefully. As mentioned previously, there is no interaction between the spins in this limit; it follows that the spins are independent of one another. When you take the average of the product of two independent spins, the result is the product of the two average values. That is,

$$\langle s_1 s_2 \rangle = \langle s_1 \rangle \langle s_2 \rangle.$$

In this system there is no magnetic field, and hence the spins are equally likely to be +1 or −1. As a result, the average value of each spin is 0, and

$$\langle s_1 s_2 \rangle = \langle s_1 \rangle \langle s_2 \rangle = 0.$$

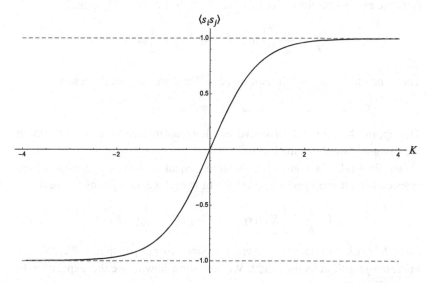

Figure 2.7 Spin–spin correlation $\langle s_1 s_2 \rangle$ as a function of the coupling K for a two-spin system.

In the other two limits, where $K \to +\infty$ and $K \to -\infty$, the spins are still equally likely to be $+1$ or -1. That hasn't changed. What has changed, however, is that the spins are correlated – when one spin is $+1$, the other is equal to $+1$ or -1, depending on the sign of K. As a result, the average value of their *product* is nonzero.

To this point, we've obtained $\langle s_1 s_2 \rangle$ by a direct summation over spin states. This is always a valid approach, though sometimes it can be rather difficult. Thus, it's also worthwhile to apply the simpler method developed in the previous section, where an average value is given by the derivative of the free energy f with respect to the conjugate variable. In this case, we have

$$f = \frac{1}{N} \ln Z = \frac{1}{2} \ln \sum_{\{s_i\}} e^{K s_1 s_2}.$$

Clearly, the conjugate variable to $s_1 s_2$ is the coupling K. Taking the corresponding derivative yields

$$\frac{\partial f}{\partial K} = \frac{1}{2}\left[\frac{1}{Z} \frac{\partial}{\partial K} \sum_{\{s_i\}} e^{K s_1 s_2} \right] = \frac{1}{2}\left[\frac{1}{Z} \sum_{\{s_i\}} s_1 s_2 e^{K s_1 s_2} \right] = \frac{1}{2}\langle s_1 s_2 \rangle.$$

We see that the derivative gives the desired average value, up to a factor of one half – this is to be expected, because f is the reduced free energy per site, and this system has two sites but only one bond. We can evaluate the derivative from the expression for f given earlier, $f = \frac{1}{2}\ln 2(e^K + e^{-K})$, to find

$$\frac{\partial f}{\partial K} = \frac{\partial}{\partial K}\left(\frac{1}{2}\ln 2(e^K + e^{-K}) \right) = \frac{1}{2}\tanh K.$$

The same factor of one half appears here. Thus, we confirm the result

$$\langle s_1 s_2 \rangle = \tanh K.$$

This approach, using the derivative, is often a useful alternative to performing the straightforward summation.

One final point is that $\partial f / \partial K$ is almost equal to the average value of the reduced Hamiltonian per spin, $-\beta H / N$. In fact, if we multiply by K, we find

$$K\frac{\partial f}{\partial K} = \frac{1}{2}K\langle s_1 s_2 \rangle = \frac{1}{N}\langle -\beta H \rangle = \frac{1}{N}(-\beta E). \tag{2.14}$$

Thus, $K \partial f / \partial K$ is equal to the average reduced energy per spin, $(-\beta E)/N$ – an interesting result in its own right. We now show how to use this expression to obtain the entropy.

The Entropy

Recall that the free energy F of an Ising system is equal to the average energy E minus the temperature T times the entropy S:

$$F = E - TS.$$

To obtain the reduced free energy per spin, f, we multiply by $-\beta = -1/k_B T$ and divide by the number of spins, N. The result is

$$f = \frac{1}{N}(-\beta F) = \frac{1}{N}(-\beta E) + \frac{1}{N}(\beta TS) = \frac{1}{N}\langle -\beta H \rangle + \frac{S}{Nk_B}.$$

Using the expression derived previously for the average energy, we can rewrite this as

$$f = K\frac{\partial f}{\partial K} + \frac{S}{Nk_B}.$$

Rearranging, and writing $s = S/Nk_B$ to denote the reduced entropy per spin, we find

$$s = f - K\frac{\partial f}{\partial K}. \tag{2.15}$$

In this particular case, with the free energy given by $f = \frac{1}{2}\ln 2(e^K + e^{-K})$, we find

$$s = \frac{1}{2}(\ln[4\cosh K] - K\tanh K).$$

This result is plotted in Figure 2.8. Notice that the high-temperature ($K \to 0$) limit is $s = \ln 2$, and the low-temperature limit ($K \to \infty$) is $s = \frac{1}{2}\ln 2$, as expected.

The Specific Heat

Now that we have an expression for the entropy, we can use it to calculate the reduced specific heat,

$$c = C/Nk_B.$$

Recall that the specific heat C is the amount of heat ($T\partial S$) required to produce a given increase in the temperature (∂T). As such, it is a quantity that is readily measured in the lab.

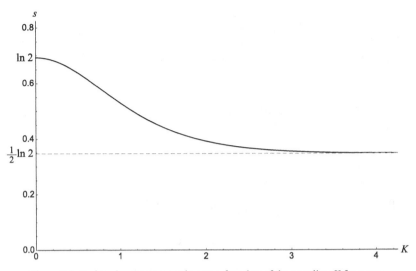

Figure 2.8 Reduced entropy per spin s as a function of the coupling K for a two-spin system.

Let's derive an expression for c using our previous results. We start by taking a temperature derivative of the entropy in Equation 2.15 and multiplying by T:

$$c = T\frac{\partial s}{\partial T} = T\frac{\partial}{\partial T}\left(f - K\frac{\partial f}{\partial K}\right) = T\left(\frac{\partial f}{\partial K}\frac{\partial K}{\partial T} - \frac{\partial K}{\partial T}\frac{\partial f}{\partial K} - K\frac{\partial^2 f}{\partial K^2}\frac{\partial K}{\partial T}\right).$$

Next, recall that $K = J/k_B T$. It follows that

$$\frac{\partial K}{\partial T} = -\frac{J}{k_B T^2} = -\frac{K}{T}.$$

Making this substitution, we find

$$c = T\frac{\partial s}{\partial T} = T\left(\frac{\partial f}{\partial K}\left(\frac{-K}{T}\right) - \left(\frac{-K}{T}\right)\frac{\partial f}{\partial K} - K\frac{\partial^2 f}{\partial K^2}\left(\frac{-K}{T}\right)\right).$$

Simplifying, and canceling where possible, yields the surprisingly simple expression

$$c = K^2\frac{\partial^2 f}{\partial K^2}. \tag{2.16}$$

Notice that the first derivative of the free energy is related to the average energy, as in Equation 2.14, $K\partial f/\partial K = (-\beta E)/N$, and the second derivative is related to the specific heat, as advertised.

We know from Equation 2.12 that the reduced free energy can be written as a function of K as follows:

$$f = \frac{1}{2}\ln 4\cosh K.$$

Taking two derivatives with respect to K and multiplying by K^2 yields

$$c = K^2 \frac{\partial^2 f}{\partial K^2} = \frac{1}{2}\frac{K^2}{\cosh^2 K}. \tag{2.17}$$

This function is plotted in Figure 2.9.

The key feature of the specific heat is that it has a peak where the interaction energy J is on the order of the thermal energy, $k_B T$. This is a consistent feature in both finite and infinite Ising models, and in any dimension.

On an intuitive level, we can understand the appearance of a peak as follows: First, there is a finite energy gap between the ground state and the next higher energy level. Therefore, at low temperatures the thermal energy isn't insufficient to produce excitations. As a result, the energy doesn't change much with temperature, and the specific heat is essentially zero. Second, at high temperatures the energy levels are occupied randomly, and hence the average energy doesn't change with temperature. Once again, this leads to $c = 0$. Third, at

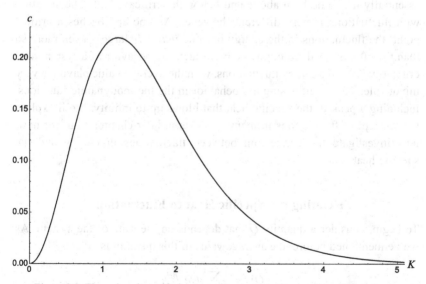

Figure 2.9 The reduced specific heat per site, c, versus the coupling, K, for a two-spin system. A peak occurs when the thermal energy is on the order of the interaction energy; that is, on the order of $K \sim 1$.

intermediate temperatures excitations are just beginning to occur, and the energy of the system depends sensitively on the temperature. This produces a peak in c.

Finally, note that the specific heat is an even function of K in this system. It follows that the plot of c for negative K is the same as in Figure 2.9 for positive K.

Fluctuations

We've learned a lot about the average energy of a system, and how it changes with temperature. It's important to remember, however, that the average energy is just that – an average. At any given time, the energy can be greater than or less than the average; that is, the energy can *fluctuate* about its average. What remains constant in the system is the temperature, not the energy.

To gain an appreciation for the importance of these fluctuations, and the reason for our interest, consider the following analogy. Suppose a hurricane is passing directly over a city. On one side of town the wind speed might be 50 m/s to the north, and on the other side of town 50 m/s to the south. Now, on average, the wind velocity is zero, which by itself would indicate a calm, peaceful day. But clearly, the average doesn't do justice to the actual conditions on the ground – the extremes are much more important in this case.

The situation is similar in systems near critical points. As one approaches a critical point, the average energy varies smoothly, with a value that is essentially the same both above and below the critical point. The situation with fluctuations is very different, however. As one approaches a critical point, the fluctuations in the energy become *infinitely* large – even more so than the 50 m/s winds in our analogy. In fact, the behavior of a system near criticality is *dominated* by fluctuations, with the average value playing a very minor role. The result is singular behavior in the thermodynamic functions, including a peak in the specific heat that blows up to infinity. We'll explore the concept of fluctuations in many contexts in later chapters; but for now, let's investigate the connection between fluctuations in energy and the specific heat.

Relating the Specific Heat to Fluctuations

To begin, consider a quantity Q that depends on the state of the system. As we've mentioned before, the average value of this quantity is

$$\langle Q \rangle = \frac{1}{Z} \sum_{\{s_i\}} Q e^{-\beta H}.$$

Now, how would we calculate the fluctuation of Q about this average value? A first guess might be to consider the difference between Q and its average value, $Q - \langle Q \rangle$, and average this difference over all states. This simply gives zero, however:

$$\langle Q - \langle Q \rangle \rangle = \langle Q \rangle - \langle Q \rangle = 0.$$

Clearly this isn't very helpful; the positive and negative fluctuations have canceled one another.

To correct this problem, let's first square $Q - \langle Q \rangle$, and then take the average. In this way, all fluctuations – positive and negative – contribute equally. Thus, consider the following average:

$$\langle (Q - \langle Q \rangle)^2 \rangle.$$

Expanding the square, and canceling where possible, we find

$$\langle (Q - \langle Q \rangle)^2 \rangle = \langle Q^2 - 2Q\langle Q \rangle + \langle Q \rangle^2 \rangle = \langle Q^2 \rangle - \langle Q \rangle^2.$$

Thus, we'll define the fluctuation of a quantity to be the average of its square minus the square of its average, $\langle Q^2 \rangle - \langle Q \rangle^2$. In the special case where Q is the reduced Hamiltonian, the fluctuation is $\langle (-\beta H)^2 \rangle - \langle (-\beta H) \rangle^2$.

Now, it turns out that the *fluctuation* in energy is related directly to the *second* derivative of the free energy. To see how, recall that the first derivative of the free energy with respect to K is

$$\frac{\partial f}{\partial K} = \frac{\partial}{\partial K}\left(\frac{1}{N}\ln Z\right) = \frac{1}{N}\frac{1}{Z}\left(\frac{\partial Z}{\partial K}\right).$$

Similarly, the second derivative is

$$\frac{\partial^2 f}{\partial K^2} = \frac{\partial}{\partial K}\left(\frac{\partial f}{\partial K}\right) = \frac{\partial}{\partial K}\left(\frac{1}{N}\frac{1}{Z}\left(\frac{\partial Z}{\partial K}\right)\right) = \frac{1}{N}\frac{\partial}{\partial K}\left(\frac{1}{Z}\left(\frac{\partial Z}{\partial K}\right)\right).$$

The derivative of the product of terms $1/Z$ and $\partial Z/\partial K$ can be evaluated with the chain rule to yield the following:

$$\frac{\partial^2 f}{\partial K^2} = \frac{1}{N}\left[\frac{1}{Z}\frac{\partial^2 Z}{\partial K^2} - \left(\frac{1}{Z}\frac{\partial Z}{\partial K}\right)^2\right].$$

Noting that

$$\frac{1}{Z}\frac{\partial Z}{\partial K} = \frac{1}{Z}\sum_{\{s_i\}} s_1 s_2 e^{K s_1 s_2} = \langle s_1 s_2 \rangle$$

and

$$\frac{1}{Z}\frac{\partial^2 Z}{\partial K^2} = \frac{1}{Z}\sum_{\{s_i\}} (s_1 s_2)^2 e^{K s_1 s_2} = \langle (s_1 s_2)^2 \rangle,$$

we have the result that

$$\frac{\partial^2 f}{\partial K^2} = \frac{1}{N}[\langle (s_1 s_2)^2 \rangle - \langle s_1 s_2 \rangle^2].$$

Thus, one way to look at the second derivative of the free energy is as the fluctuation of the product of spins.

If we now multiply by K^2, we obtain the reduced specific heat:

$$c = K^2 \frac{\partial^2 f}{\partial K^2} = \frac{1}{N}[\langle (K s_1 s_2)^2 \rangle - \langle K s_1 s_2 \rangle^2].$$

In this system, $K s_1 s_2$ is the reduced Hamiltonian, $-\beta H$, and hence the reduced specific heat is simply the fluctuation in the energy of the system:

$$c = K^2 \frac{\partial^2 f}{\partial K^2} = \frac{1}{N}[\langle (-\beta H)^2 \rangle - \langle (-\beta H) \rangle^2]. \qquad (2.18)$$

This connection between the specific heat and energy fluctuations is quite general, and of fundamental significance.

Let's summarize by pointing out that we have two very different expressions for the specific heat. We'll display them here for comparison:

$$c = T \frac{\partial s}{\partial T}$$
$$c = \frac{1}{N}[\langle (-\beta H)^2 \rangle - \langle (-\beta H) \rangle^2]. \qquad (2.19)$$

In the first expression, the specific heat is given in terms of thermodynamic quantities; namely, as the heat corresponding to a given temperature change. This is the common interpretation of the specific heat in undergraduate thermodynamic courses. The second expression gives a completely different understanding of the specific heat – as the fluctuation in the energy of the system. Thus, when we measure the specific heat, we are in effect measuring energy fluctuations. When the specific heat diverges to infinity, as it does in the two-dimensional Ising model, for example, we know that fluctuations are unbounded.

2.3 Three Spins

We are ultimately interested in studying infinite systems, because that's where true phase transitions occur; but still, it's worthwhile to proceed slowly and develop more insight into Ising systems as we go. In going from one to two spins, we learned about spin–spin correlations and the connections between energy fluctuations and the specific heat. In this section, we consider three-spin systems, which will give us insight into various aspects of symmetry in Ising systems, as well as the surprisingly important role played by topology.

Free Boundary Conditions

Let's start by adding one spin to the system in Figure 2.5. The result is a three-spin system, $N = 3$, with three spins in a line, as in Figure 2.10. The spins are labeled 1, 2, and 3, counting from left to right, and there are nearest-neighbor interactions K between spins 1 and 2 and spins 2 and 3, as indicated. The group of spins simply terminates at spin 1 and spin 3, which we refer to as *free boundary conditions*. It follows that the reduced Hamiltonian is

$$-\beta H = K(s_1 s_2 + s_2 s_3).$$

As before, $K > 0$ is a ferromagnetic interaction, and $K < 0$ is an antiferromagnetic interaction.

With three spins, the number of states is $2^3 = 8$. There is no magnetic field in this system, and hence states in which all the spins are reversed in sign – from $+1$ to -1 and from -1 to $+1$ – give the same value for the reduced Hamiltonian. Thus, there are only four independent spin states, or configurations. These configurations and the corresponding value of the reduced Hamiltonian are given in the following table:

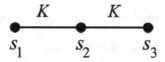

Figure 2.10 A three-spin system with free boundary conditions.

Spin configurations	+++	++−	−++	+−+
	−−−	−−+	+−−	−+−
Reduced Hamiltonian	2K	0	0	−2K

This table shows not only the up–down symmetry in flipping all spins, but also the left–right symmetry about the center of the system.

The partition function for this system can now be read directly from the table:

$$Z = \sum_{\{s_i\}} e^{-\beta H} = \sum_{\{s_i\}} e^{K(s_1 s_2 + s_2 s_3)} = 2(e^{2K} + 2 + e^{-2K}).$$

The corresponding reduced free energy per spin is

$$f = \frac{1}{N}\ln Z = \frac{1}{3}\ln[2(e^{2K} + 2 + e^{-2K})] = \frac{1}{3}\ln[8(\cosh K)^2].$$

Notice that the free energy is even in K, yet another form of symmetry in this system.

As before, let's take a look at the limits of the free energy. First, the zero-coupling, infinite-temperature limit:

$$K \to 0 \qquad f \to \frac{1}{3}\ln 8 = \ln 2 \qquad \text{all entropy, no energy}$$
$$T \to \infty.$$

This limit is always a good check to make sure that no obvious mistakes have been made in writing our expression for f.

Next, we examine the infinite-coupling, zero-temperature limit:

$$K \to \infty \qquad f \to \frac{1}{3}\ln 2 + \frac{2}{3}K \qquad \text{ground-state entropy and energy}$$
$$T \to 0.$$

As with the two-spin system, there are two ground states, $+++$ and $---$. In this case, however, there are three spins, $N = 3$, and hence the ground-state entropy is $\frac{1}{3}\ln 2$. In the limit that $N \to \infty$ the ground-state entropy vanishes, as long as there are only a finite number of ground states.

The reduced energy of the ground states is $2K$, which gives a reduced energy of $\frac{2}{3}K$ per spin. The reduced free energy is plotted in Figure 2.11, for both positive and negative K. Notice that $f \to \frac{1}{3}\ln 2 - \frac{2}{3}K$ in the zero-temperature antiferromagnetic limit of $K \to -\infty$.

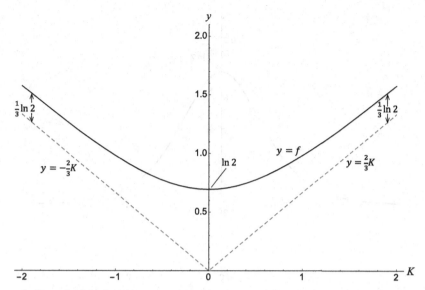

Figure 2.11 Reduced free energy per spin f versus the coupling K for a three-spin system with free boundary conditions. The left and right dashed lines represent $y = -2K/3$ and $y = 2K/3$, respectively.

Let's also take a look at the reduced entropy per spin, $s = f - K \partial f / \partial K$. Using our previous results, we find

$$s = \frac{1}{3} \ln[8(\cosh K)^2] - \frac{2}{3} K \tanh K.$$

This result is plotted in Figure 2.12. The entropy has its greatest value, $\ln 2$, at $K = 0$, which corresponds to infinite temperature. It drops to its ground-state value, $\frac{1}{3} \ln 2$, for $K \to \pm \infty$.

Finally, consider the reduced specific heat per spin, $c = K^2 \partial^2 f / \partial K^2$, for this system. Taking the appropriate derivatives of the reduced free energy, we find

$$c = \frac{2}{3} \frac{K^2}{(\cosh K)^2}.$$

Figure 2.13 shows c for both positive and negative K; notice the symmetry about $K = 0$. In addition, note that the specific heat peaks when the thermal energy is comparable to the interaction energy.

Referring to Equation 2.17, we see that the reduced specific heat in this case is practically the same as it was for two spins. We'll discuss the reasons for this in the next chapter.

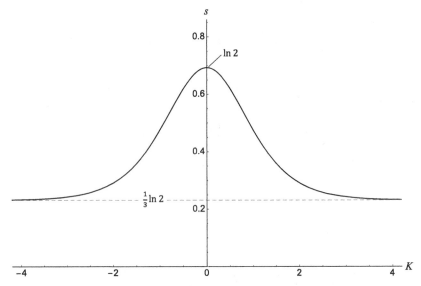

Figure 2.12 Reduced entropy per spin for the three-spin system with free boundary conditions.

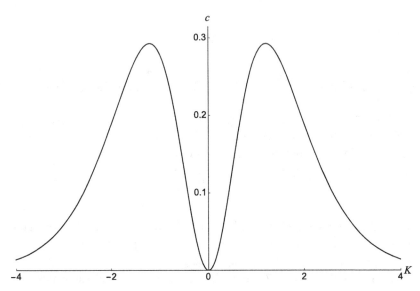

Figure 2.13 Reduced specific heat per spin c versus coupling K for a three-spin system with free boundary conditions.

Periodic Boundary Conditions

Let's consider the same three spins, but this time with boundary conditions that have no end; that is, *periodic boundary conditions*. What does this mean? Well, it means that when we get to the end of the line, at s_3, we include an additional interaction between it and the first spin in the line, s_1. It's a bit like old-time computer games, where a character exits one side of the screen and reappears immediately on the other side. We can visualize the situation by imagining our line of spins wrapped on a ring, as indicated in Figure 2.14.

Alternatively, when there are just three spins, as in this case, we can arrange them in a symmetric group. For example, we can place the spins at the vertices of an equilateral triangle, as in Figure 2.15. A situation like this is a bit easier to draw.

In either case, what's really important is the reduced Hamiltonian, which differs from the one used for free boundary conditions by the inclusion of an extra interaction, Ks_3s_1. We can write the reduced Hamiltonian as follows:

$$-\beta H = K(s_1s_2 + s_2s_3 + s_3s_1).$$

This is a simple change, but it has rather profound implications, as we shall see. In most cases, periodic boundary conditions, which wrap around back to the beginning, are a better approximation to an infinite lattice than are free boundary conditions, which simply truncate the lattice. There are exceptions to this rule, however, as we shall see in the next chapter.

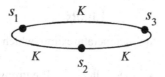

Figure 2.14 A three-spin system with periodic boundary conditions, pictured as spins on a ring.

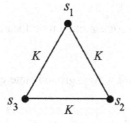

Figure 2.15 A three-spin system with periodic boundary conditions can also be pictured as three spins at the vertices of an equilateral triangle.

To obtain the partition function, we examine the eight spin configurations of this system and the corresponding values of the reduced Hamiltonian.

Spin configurations	$+$	$-$	$+$	$-$
	$++$	$++$	$--$	$--$
Reduced Hamiltonian	$3K$	$-K$	$-K$	$3K$

Note that there is just one configuration with all spins $+$, three configurations with one spin $-$, three configurations with two spins $-$, and one configuration with all three spins. $-$.

Summing over these configurations yields

$$Z = \sum_{\{s_i\}} e^{-\beta H} = \sum_{\{s_i\}} e^{K(s_1s_2+s_2s_3+s_3s_1)} = 2(e^{3K} + 3e^{-K}).$$

The reduced free energy per spin is

$$f = \frac{1}{N}\ln Z = \frac{1}{3}\ln[2(e^{3K} + 3e^{-K})]. \tag{2.20}$$

This time f is not an even function of K. There's an interesting reason for this, as we shall see.

Limits

We begin with the simple high-temperature limit, which is our usual first check:

$$K \to 0 \qquad f \to \frac{1}{3}\ln 8 = \ln 2 \qquad \text{all entropy, no energy}$$
$$T \to \infty.$$

In general, with N spins we have 2^N states. It follows that the reduced entropy per spin is $s = \frac{1}{N}\ln 2^N = \ln 2$.

Next, we consider the ferromagnetic ground state, $K \to +\infty$. In this limit, we find

$$K \to +\infty \quad f \to \frac{1}{3}\ln 2 + K \qquad \text{ground-state entropy and energy}$$
$$T \to 0.$$

The ground states have a reduced energy of $3K$, and there are three spins in the system. Hence, the reduced energy per spin for the ground states is K. There are two ground states – all spins up and all spins down – and hence $\frac{1}{3}\ln 2$ is the reduced entropy per spin. These observations account for the two terms in the limit.

Frustration

In the other systems we've studied so far, there has been no need to investigate the limit $K \to -\infty$ in addition to $K \to +\infty$, because the free energy has always been an even function of K. That is not the case here. The limit $K \to -\infty$ will yield not only new results for this system, but interesting new physics as well.

Consider, then, the antiferromagnetic ground state, $K \to -\infty$. In this limit we find

$$K \to -\infty \qquad f \to \frac{1}{3}\ln 2 + \frac{1}{3}\ln 3 - \frac{1}{3}K \qquad \text{ground-state}$$
$$T \to 0. \qquad\qquad\qquad\qquad\qquad\qquad \text{entropy and energy}$$

This is certainly different from the ferromagnetic limit, $K \to +\infty$, as expected. First, there is now a multiplicative factor of $\frac{1}{3}$ in front of $-K$ in the energy term; second, there is an additional entropy contribution of $\frac{1}{3}\ln 3$.

To account for these differences, note that the behavior associated with ferromagnetic and antiferromagnetic states is quite different for free and periodic boundary conditions. Starting with the case of free boundary conditions, as shown in the top of Figure 2.16, note that one of the ferromagnetic ground states for $K \to +\infty$ is $+ + +$. Each of the two bonds in this state is "satisfied" in the sense that each pair of interacting spins has the same sign, as expected for a ferromagnetic interaction. The reduced Hamiltonian for this ground state is $2K$. In the top of Figure 2.16 (b) we see one of the antiferromagnetic ground states in the limit $K \to -\infty$; namely, $+ - +$. Again, each of the bonds is satisfied because the interacting spins now have opposite (antiferromagnetic) signs. The reduced Hamiltonian is $-2K$, just what we had in the ferromagnetic case, but with $K \to -K$; this is why the free energy with free boundary conditions is an even function in K.

Let's turn now to the case of periodic boundary conditions, shown in the bottom of Figure 2.16. Consider one of the two ferromagnetic ground states, $+ + +$. All the bonds are satisfied, and the reduced energy is $3K$; that is, K per spin. The situation is quite different with the antiferromagnetic ground states, however. There are six of these ground states, three with one minus spin, $- + +, + - +, + + -$, and three spin-reversed states with two minus spins, $+ - -, - + -, - - +$. Each of these ground

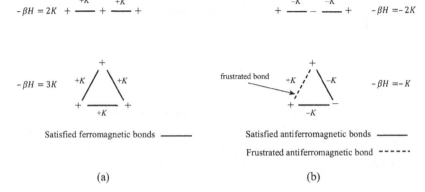

Figure 2.16 (a) In the three-spin system with ferromagnetic interactions, all bonds are satisfied for both free (top) and periodic (bottom) boundary conditions. (b) With antiferromagnetic interactions, one bond is always frustrated with periodic boundary conditions.

states, like the one shown in the bottom of Figure 2.16 (b), has two bonds that satisfy the antiferromagnetic condition (interacting spins with opposite signs), but one bond that is unsatisfied – with interacting spins of the same sign. This situation simply can't be avoided because of the topology of the equilateral triangle. We say that this system – which can't satisfy all of its bonds – is *frustrated*.

The reduced Hamiltonian for each of these frustrated states is $-K$; that is $-\frac{1}{3}K$ per spin. This accounts for the energy term in the $K \to -\infty$ limit. The ground-state entropy term corresponds to six states in a system of three spins:

$$\frac{1}{3}\ln 6.$$

Writing the six as two times three, we can express the ground-state entropy as

$$\frac{1}{3}\ln 2 + \frac{1}{3}\ln 3.$$

The first term represents the entropy of spin reversal, and the second term represents the additional entropy of frustration.

Thermodynamic Functions

In Figure 2.17 we plot the reduced free energy per spin f for three spins with periodic boundary conditions. We plot this result as a function of K, including

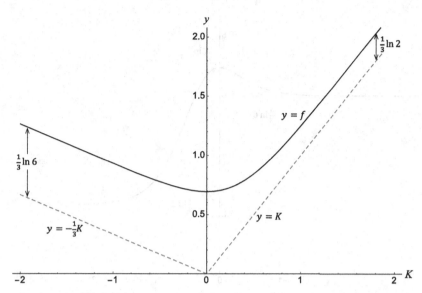

Figure 2.17 Reduced free energy per spin (solid curve) for the three-spin system with periodic boundary conditions. Notice the asymmetry about $K = 0$. The left and right dashed lines represent $y = -K/3$ and $y = K$, respectively.

both positive and negative values. We see that f is not symmetric about $K = 0$, as expected. Note that $f = \ln 2$ for $K = 0$, and that it approaches the line $y = K$ for $K \to +\infty$ and the line $y = -\dfrac{1}{3}K$ for $K \to -\infty$, with appropriate offsets. The large offset for negative K is due to frustration.

Next, let's take a look at the reduced entropy per spin, s, for this system. Recall that $s = f - K\partial f/\partial K$; therefore, we find

$$s = \frac{1}{3}\ln[2(e^{3K} + 3e^{-K})] - K\left(\frac{e^{3K} - e^{-K}}{e^{3K} + 3e^{-K}}\right).$$

This result is plotted in Figure 2.18. Again, we see the asymmetry about $K = 0$, and the large entropy contribution from frustration for negative K.

Finally, we calculate the reduced specific heat per spin, c. One definition of the specific heat is $c = T\partial s/\partial T$. From this expression, we expect a larger specific heat for positive K, as opposed to negative values of K, because of the large change in entropy there. We'll check this conclusion by calculating the specific heat with K^2 times the second derivative of $f(K)$:

$$c = K^2\frac{\partial^2 f}{\partial K^2} = \frac{4K^2}{[\sinh(2K) - 2\cosh(2K)]^2}.$$

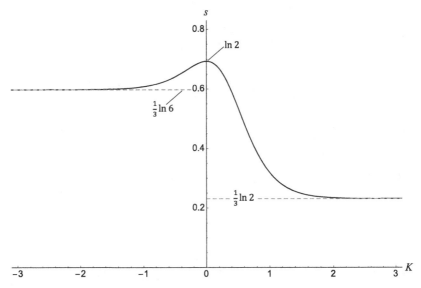

Figure 2.18 Reduced entropy per spin s for the three-spin system with periodic boundary conditions. Notice the large entropy due to frustration for negative K.

This result is plotted in Figure 2.19. We see immediately the asymmetry in c, and the larger values for positive K. There is more entropy due to frustration for negative values of K, as we see in Figure 2.18, but the specific heat depends on the *rate of change* of the entropy, and this is larger for positive K.

2.4 $N \times N$ Systems

We know that the infinite two-dimensional Ising model has a true phase transition. In this section, we make approximations to the infinite lattice by considering larger and larger finite lattices. These lattices will be an $N \times N$ section of the infinite lattice; that is, N lattice sites in the horizontal direction and N lattice sites in the vertical direction. We can only go so far with this approach before the calculations become too burdensome, but we will gain valuable insight into the infinite system from the finite cases that are tractable.

To begin, let's consider the question of how to cover the two-dimensional square lattice in a way that counts every nearest-neighbor interaction once, and only once. This is accomplished by assigning one vertical and one horizontal nearest-neighbor coupling to each spin. This is indicated in Figure 2.20. On the left, we show a given spin interacting with the spin above it, and the spin to its

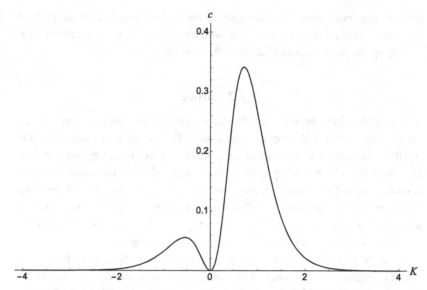

Figure 2.19 Reduced specific heat per spin c for the three-spin system with periodic boundary conditions. The large change in entropy for positive K in Figure 2.18 accounts for the large peak to the right of the origin.

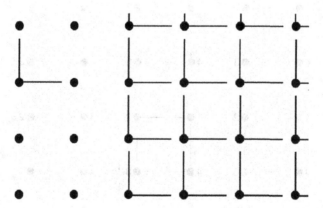

Figure 2.20 One vertical and one horizontal interaction per spin accounts for all the nearest-neighbor interactions on a square lattice.

right, with the interactions indicated by the solid lines. On the right, we show a group of spins with their interactions, showing that these two interactions per spin account for all the nearest-neighbor pairs. Thus, there are two nearest-neighbor interactions per site on the square lattice. This assignment rule – one vertical interaction, and one horizontal interaction – will be helpful in setting up the finite lattices studied in this section.

2 × 2 Lattice

The simplest finite lattice with the basic features of a square lattice is an elementary square with two sites on a side. Before we start calculating the partition function for four spins in a square array, however, we want to introduce periodic boundary conditions, which will help the finite lattice to resemble an infinite lattice more closely. To extend the idea of periodic boundary conditions from a one-dimensional system, as in the previous section, to a two-dimensional square lattice, we imagine making copies of our system and tiling the plane with them. This is illustrated in Figure 2.21.

Next, we go to each of our four spins and give it a vertical and a horizontal nearest-neighbor interaction. For example, spin 1 in Figure 2.21 interacts with the spin above it – which is spin 4, as per our periodic boundary conditions – and with the spin to the right, spin 2. Similarly, spin 2 interacts with spin 3 and spin 1, spin 3 interacts with spin 2 and spin 4, and spin 4 interacts with spin 1

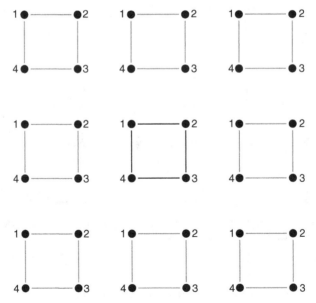

Figure 2.21 Periodic boundary conditions for a 2 × 2 square lattice correspond to replicating the 2 × 2 square to form an infinite lattice. The original 2 × 2 lattice is shown in the center in bold; replicas are shown ghosted.

and spin 3. The net result is that we have simply doubled the interactions on each side of the square from K to $2K$. Our reduced Hamiltonian is

$$-\beta H = 2K(s_1 s_2 + s_2 s_3 + s_3 s_4 + s_4 s_1).$$

Using periodic boundary conditions, we have correctly assigned two nearest-neighbor interactions, K, per site.

The partition function for this system consists of a sum over the $2^4 = 16$ spin states. This sum can be carried out by hand, with the result

$$Z = 2e^{8K} + 12 + 2e^{-8K}.$$

The reduced free energy per site follows immediately:

$$f = \frac{1}{N} \ln Z = \frac{1}{4} \ln(2e^{8K} + 12 + 2e^{-8K}).$$

Notice that $f = \ln 2$ in the limit $K \to 0$, as expected.

The reduced specific heat per site for this system is

$$c = K^2 \frac{\partial^2 f}{\partial K^2} = \frac{16K^2(1 + 3\cosh 8\,K)}{(3 + \cosh 8\,K)^2}.$$

Figure 2.22 shows a plot of c as a function of K. Notice that it has a peak when the thermal energy is roughly double the interaction energy. In fact, we know that the infinite lattice has a peak that goes to infinity at $K = 0.44069 \ldots$, the location of which is indicated in Figure 2.22 by the vertical dashed line. Thus, even with this simple system, approximating an infinite lattice by only four sites, we get the result that the specific heat peaks in roughly the correct location.

Larger Finite Lattices

Let's apply this finite-lattice approximation scheme to larger lattices. In Figure 2.23 we show the 3×3 lattice, periodically extended in the same way we extended the 2×2 lattice earlier. The reduced Hamiltonian for this system, with one vertical and one horizontal interaction per spin, is

$$\begin{aligned}
-\beta H = \; & K\big(s_1(s_7 + s_2) + s_2(s_8 + s_3) + s_3(s_9 + s_1) + s_4(s_1 + s_5) \\
& + s_5(s_2 + s_6) + s_6(s_3 + s_4) + s_7(s_4 + s_8) + s_8(s_5 + s_9) \\
& + s_9(s_6 + s_7)\big).
\end{aligned}$$

The corresponding partition function, with $2^9 = 524$ states, is a bit messy to calculate – you will surely want to use a computer. The same idea can be extended to the 4×4, 5×5, and 6×6 lattices.

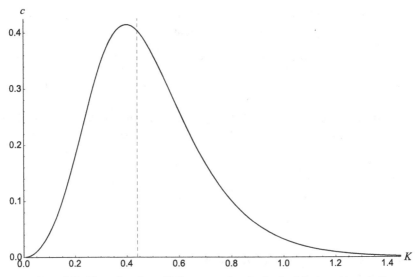

Figure 2.22 The reduced specific heat per spin c for the 2×2 lattice with periodic boundary conditions. The vertical dashed line shows the location of the peak for the infinite-system specific heat.

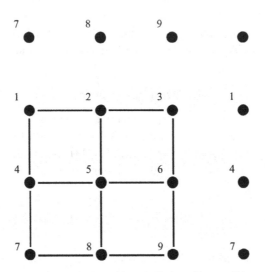

Figure 2.23 A 3×3 square lattice with periodic boundary conditions.

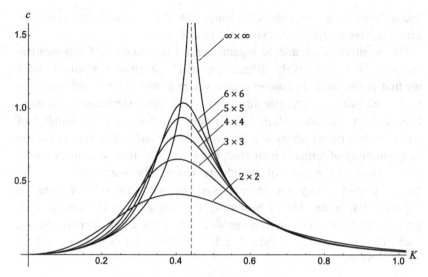

Figure 2.24 Reduced specific heat per spin c for finite lattices and for the infinite lattice. From bottom to top we have the following lattices: 2×2, 3×3, 4×4, 5×5, 6×6, $\infty \times \infty$.

In each of these cases, it is straightforward to have the computer calculate the reduced specific heat per spin using

$$c = K^2 \frac{\partial^2 f}{\partial K^2}.$$

The results for the 2×2 through 6×6 lattices are plotted in Figure 2.24. We also include for comparison the exact reduced specific heat per spin for an $\infty \times \infty$ lattice from Onsager's solution, to be covered in Chapter 5. This is the curve that diverges to infinity at the location of the dashed line, $K = 0.44069 \ldots$. Notice that the results for the finite lattices reach higher and higher peaks as the size of the system increases, and that the location of the peaks more and more closely approximates the exact location.

Singularities and the Size of a System

This is a good place to point out that an infinite system is needed to produce a singular specific heat that diverges to infinity, like the infinite-lattice case in Figure 2.24. After all, the partition function of a finite system is a finite sum of Boltzmann factors, each of which is a well-behaved, analytic exponential. Any finite sum of exponentials is still an analytic function, and taking a logarithm and differentiating twice doesn't change that fact. So, finite systems must certainly have analytic thermodynamic functions.

On the other hand, adding together an *infinite* number of exponentials can produce a qualitatively different result. The situation is similar to what we find in the study of Fourier series, where a sum of sines and cosines – nice, well-behaved, analytic functions – can approximate any function, even one that is nonanalytic, like a step function. A finite number of terms in the Fourier series will produce a better and better approximation as the number of terms is increased; but a true step function, for example, occurs only in the limit of an infinite series. The same holds true for Ising systems. Only an infinite system, with an infinite sum of exponentials, is capable of producing the singularities characteristic of a phase transition. Finite systems produce better and better approximations, each of which is nicely analytic, but the infinite system is still essential for singular behavior.

2.5 Problems

2.1 Is the ground-state entropy of the one-spin system in Figure 2.1 greater than, less than, or equal to the ground-state entropy of the two-spin system in Figure 2.5? Explain.

2.2 Three independent spins are each acted on by a magnetic field h, as in Figure 2.4 with $N = 3$. (a) Sum over the two configurations on any one of the spins to obtain its partition function, Z_1. (b) Sum over the eight configurations of the entire system to obtain the total partition function of the system, Z_{total}. (c) Raise your result for Z_1 to the third power to show that $Z_{total} = (Z_1)^3$.

2.3 Consider the following series of numbers: $S = 0, 1, 2, 3$. (a) Is $\langle S \rangle^2$ greater than, less than, or equal to $\langle S^2 \rangle$? Explain. (b) Calculate $\langle S \rangle^2$. (c) Calculate $\langle S^2 \rangle$. (d) Calculate $\langle S^2 \rangle - \langle S \rangle^2$.

2.4 Consider the following two series of numbers: $S_1 = 1, 2, 3$ and $S_2 = 1.9, 2, 2.1$. (a) Is $\langle S_1 \rangle$ greater than, less than, or equal to $\langle S_2 \rangle$? Explain. (b) Is $\langle S_1^2 \rangle - \langle S_1 \rangle^2$ greater than, less than, or equal to $\langle S_2^2 \rangle - \langle S_2 \rangle^2$? Explain. (c) Calculate $\langle S_1^2 \rangle - \langle S_1 \rangle^2$. (d) Calculate $\langle S_2^2 \rangle - \langle S_2 \rangle^2$.

2.5 Suppose the two-spin Ising system in Figure 2.5 has both a nearest-neighbor coupling K and an external magnetic field h. The reduced Hamiltonian is

$$-\beta H = Ks_1s_2 + h(s_1 + s_2).$$

(a) Calculate the partition function for this system. (b) Calculate the reduced free energy per spin.

2.6 Add an external magnetic field to the three-spin system with periodic boundary conditions shown in Figure 2.15. (a) What is the Hamiltonian for this new system? (b) Calculate the partition function for this new system.

2.7 Add an external magnetic field to the three-spin system with free boundary conditions shown in Figure 2.10. (a) What is the Hamiltonian for this new system? (b) Consider the following three prospective partition functions:

$$Z_1 = e^{4K}(e^{3h} + e^{-3h}) + e^{-4K}(e^h + e^{-h}) + 2(e^h + e^{-h})$$

$$Z_2 = e^{2K}(e^{3h} + e^{-3h}) + e^{-2K}(e^h + e^{-h}) + 2(e^h + e^{-h})$$

$$Z_3 = 2e^{2K}(e^{3h} + e^{-3h}) + 2e^{-2K}(e^h + e^{-h}) + 4(e^h + e^{-h}).$$

Identify the one that is correct for this system, and give a reason why each of the others is incorrect.

2.8 Figure 2.25 shows three curves (1, 2, 3) representing – in no particular order – the free energy, entropy, and specific heat of a finite Ising system plotted as a function of K. (a) Which curve is the reduced specific heat per site? Explain. (b) Which curve is the reduced free energy per site? Explain. (c) Which curve is the reduced entropy per site? Explain.

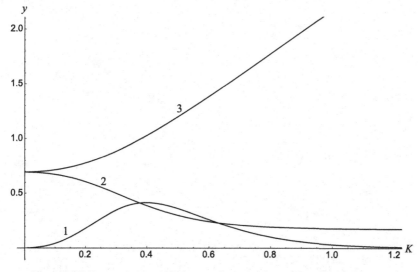

Figure 2.25 Free energy, specific heat, and entropy for an Ising system.

2.9 Add an external magnetic field h to the 2×2 system in Figure 2.21. (a) Write the Hamiltonian for this system. (b) Calculate the reduced free energy per spin, $f(K, h)$.

2.10 What is the ground-state $(K \to +\infty)$ reduced entropy per spin for the 3×3 system shown in Figure 2.23?

3

Partial Summations and Effective Interactions

A useful way to solve a complex problem – whether in physics, mathematics, or life in general – is to break it down into smaller pieces that can be handled more easily. This is especially true of the Ising model. In this chapter, we investigate various partial-summation techniques in which a subset of Ising spins is summed over to produce new, effective couplings among the remaining spins. These methods are useful in their own right, and are even more important when used as part of the renormalization-group techniques to be covered in Chapter 7.

3.1 Preserving the Partition Function

The basic requirement of all partial-summation techniques is to *preserve the partition function*. When the partition function is preserved, the free energy, and all other thermodynamic functions, are preserved as well. Thus, preserving the partition function ensures that the thermodynamics of the system is unchanged.

Ising Partial Summation

Suppose we would like to calculate the partition function, Z, for an Ising system with a reduced Hamiltonian $-\beta H(\{s_i\})$. This can be expressed as follows:

$$Z = \sum_{\{s_i\}} e^{-\beta H(\{s_i\})}.$$

It is the calculation of this partition function that is our primary goal.

Now, imagine dividing the spins s_i into two groups; one group we label μ_i, and the other σ_i. Suppose that, instead of summing over *all* the spins, as is usual,

65

we do a *partial summation*; that is, we sum over just one of the groups of spins – let's say the σ_i spins. Doing this partial summation generates an *effective*, or *renormalized*, reduced Hamiltonian $-\beta H'(\{\mu_i\})$ that depends only on the remaining spins, μ_i. The transformation can be written as follows:

$$e^{-\beta H'(\{\mu_i\})} = \sum_{\{\sigma_i\}} e^{-\beta H(\{\mu_i\},\{\sigma_i\})}. \tag{3.1}$$

The corresponding primed partition function, Z', expressed in terms of the renormalized Hamiltonian, is

$$Z' = \sum_{\{\mu_i\}} e^{-\beta H'(\{\mu_i\})}. \tag{3.2}$$

Combining Equations 3.1 and 3.2, we see that

$$Z' = \sum_{\{\mu_i\}} e^{-\beta H'(\{\mu_i\})} = \sum_{\{\mu_i\}} \sum_{\{\sigma_i\}} e^{-\beta H(\{\mu_i\},\{\sigma_i\})} = \sum_{\{s_i\}} e^{-\beta H(\{s_i\})} = Z.$$

Thus, a partial-summation transformation defined in this way preserves the partition function, as desired. It also breaks down a complex problem into more manageable pieces.

In the following sections, we consider a variety of systems where the partial summation in Equation 3.1 is carried out in detail.

3.2 Effective Magnetic Fields

The simplest application of a partial-summation transformation is a summation over one spin in a two-spin Ising system. This transformation will generate an effective Hamiltonian for the remaining spin. In this section, we focus on the mechanism of the transformation.

Consider the partial summation illustrated in Figure 3.1. On the left we see the original system; two spins, s_1 and s_2, with a nearest-neighbor interaction, K, between them, and a magnetic field, h, that acts on s_1. The reduced Hamiltonian is

$$-\beta H = K s_1 s_2 + h s_1.$$

Notice that there is no magnetic field acting directly on spin s_2.

The "X" on spin s_1 in Figure 3.1 indicates that *it* is the spin that is summed over in this transformation. We can think of s_1 as the σ_i spin in the discussion of the previous section. Similarly, s_2 is the spin that remains – the μ_i spin.

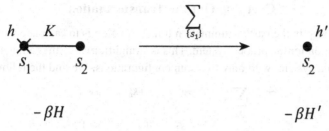

Figure 3.1 A simple partial-summation transformation. In this case, the spin s_1 is summed over, as indicated by the X. The result is an effective Hamiltonian for spin s_2.

Before proceeding with the partial summation, and the renormalized system on the right of Figure 3.1, let's pause to consider the physics of the original system. First, a magnetic field acts on spin s_1 and tends to align it in the direction of h. In addition, a positive nearest-neighbor interaction K tends to align the spin s_2 in the same direction as s_1. Thus, even though no field acts directly on s_2, it will experience an *effective* magnetic field h' due to the influence of s_1. The partial-summation transformation gives us the precise functional form of $h' = h'(K, h)$; that is, by preserving the partition function the transformation gives an h' that has the *same effect* in the renormalized system that the interactions K and h have in the original system.

Let's turn now to the renormalized system, shown on the right side of Figure 3.1. This system consists of a single spin acted on by the magnetic field h'. The Hamiltonian in this case is

$$-\beta H' = h's_2 + K_0'. \tag{3.3}$$

The term $h's_2$ is to be expected, but what is the additional term, the one labeled K_0'?

To explain this new term, we first note that the subscript zero indicates a *zero-spin* term, to distinguish it from the one-spin (h) and two-spin (K) terms of the original system. The primed zero-spin term is simply an additive energy that resets, or renormalizes, the zero level of the primed system – it's like moving an experiment to a higher floor in a building. It turns out, as we shall see later in this chapter (Section 3.5), that we have to allow for the possibility of an additive energy term to successfully preserve the partition function. In fact, K_0' is needed because it is actually a *portion* of the total free energy of the system that has been produced by the summation over a *portion* of the spins.

Carrying Out the Transformation

The first step in the partial-summation transformation is to calculate the partition function for the primed system. This is straightforward; after all, it's just a single-spin system, with only two spin configurations. We find the following:

$$Z' = \sum_{\{s_2\}} e^{-\beta H'(\{s_2\})} = e^{h' + K_0'} + e^{-h' + K_0'}$$

$$s_2 = +1 \qquad -1. \tag{3.4}$$

The first term is the contribution for $s_2 = +1$, and the second term is the contribution for $s_2 = -1$.

Next, we perform *partial* summations of the original system for each of the values of s_2. First, we set $s_2 = +1$ and sum over the values of s_1. The resulting *partial partition function*, which we label Z_+, is given by

$$Z_+ = \sum_{\{s_1, s_2 = +1\}} e^{-\beta H(\{s_1, s_2 = +1\})} = e^{K+h} + e^{-K-h}. \tag{3.5}$$

Similarly, we calculate the partial partition function Z_- that corresponds to summing over the values of s_1 with $s_2 = -1$. The result is

$$Z_- = \sum_{\{s_1, s_2 = -1\}} e^{-\beta H(\{s_1, s_2 = -1\})} = e^{-K+h} + e^{K-h}. \tag{3.6}$$

Notice that if we sum these two partial partition functions, which cover both the possible values of s_2, the result is the total partition function Z of the original system:

$$Z = Z_+ + Z_-. \tag{3.7}$$

We can now preserve the partition function, $Z = Z'$, by simply setting the $s_2 = +1$ term of Z' in Equation 3.4 equal to Z_+, and the $s_2 = -1$ term of Z' equal to Z_-. This yields

$$\begin{aligned} e^{h' + K_0'} &= Z_+ \\ e^{-h' + K_0'} &= Z_-. \end{aligned} \tag{3.8}$$

Taking the log of both sides of these equations gives

$$\begin{aligned} h' + K_0' &= \ln Z_+ \\ -h' + K_0' &= \ln Z_-. \end{aligned} \tag{3.9}$$

Subtracting the second equation from the first yields an expression for h', and adding the two equations yields an expression for K_0'. Rearranging and simplifying, we find

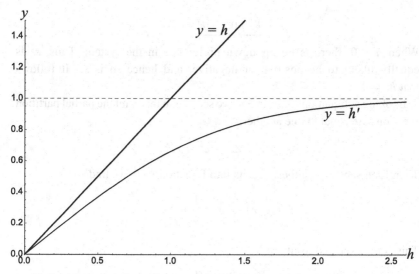

Figure 3.2 A plot of $y = h'$ versus h for $K = 1$. The 45° line is $y = h$, and the horizontal dashed line is $y = K = 1$.

$$h' = \frac{1}{2}\ln\left(\frac{Z_+}{Z_-}\right) = \frac{1}{2}\ln\left(\frac{e^{K+h} + e^{-K-h}}{e^{K-h} + e^{-K+h}}\right) \qquad (3.10)$$

$$K_0' = \frac{1}{2}\ln(Z_+ Z_-) = \frac{1}{2}\ln[(e^{K+h} + e^{-K-h})(e^{K-h} + e^{-K+h})]. \qquad (3.11)$$

We now know the effective magnetic field acting on s_2 for any given h and K, and we also know how much free energy has been generated by the partial summation.

A plot of h' versus h is given in Figure 3.2 for the specific case of $K = 1$. Notice that h' approaches a horizontal line at the value $y = K$ as $h \to \infty$. We will verify this limit later in this section. The 45° line shows that h' is always less than or equal to h, as one would expect.

A quick look at h' in Equation 3.10 shows that the transformation is invariant under the interchange of K and h. As a result, a plot of h' as a function of K for $h = 1$ looks the same as Figure 3.2.

Limits

Let's now consider some of the limits of h', both in terms of mathematics – using Equation 3.10 – as well as in terms of the corresponding physical interpretations. In the following, we use the symbol $\Rightarrow$ to stand for "implies."

$$h \to 0 \;\Rightarrow\; h' \to 0$$

When $h = 0$, there is no up–down preference in the system. Thus, s_1 is equally likely to be positive or negative, and hence so is s_2. It follows that $h' = 0$.

Looking at Equations 3.5 and 3.6, we see that in this limit the partial partition functions are equal to one another:

$$Z_+ = Z_- \to e^K + e^{-K}.$$

Therefore, substituting these results into Equation 3.10, we find

$$h' = \frac{1}{2}\ln\left(\frac{Z_+}{Z_-}\right) = \frac{1}{2}\ln(1) = 0.$$

This is the expected result.

$$K \to 0 \;\Rightarrow\; h' \to 0$$

When the coupling K between s_1 and s_2 vanishes, it follows that s_2 is no longer influenced by s_1. Therefore, s_2 is equally likely to be $+1$ or -1, regardless of what field acts on s_1. As a result, $h' = 0$.

Mathematically, we see that the partial partition functions are equal when $K \to 0$:

$$Z_+ = Z_- \to e^h + e^{-h}.$$

It follows that $h' = 0$, independent of the value of h.

$$K \to \infty \;\Rightarrow\; h' \to h$$

In this limit, s_1 and s_2 are "locked" together; that is, they always have the same value. Therefore, the magnetic field that acts on s_1 has exactly the same effect on s_2. As a result, we expect to find $h' \to h$.

From Equations 3.5 and 3.6, we see that in this limit the partial partition functions reduce to the following:

$$Z_+ \to e^{K+h}$$
$$Z_- \to e^{K-h}.$$

Using these results in Equation 3.10 yields

$$h' = \frac{1}{2}\ln\left(\frac{Z_+}{Z_-}\right) \to \frac{1}{2}\ln\left(\frac{e^{K+h}}{e^{K-h}}\right) = h.$$

This confirms our expectation.

$$h \to \infty \ \Rightarrow \ h' \to K$$

With an infinite positive magnetic field acting on s_1, it follows that s_1 is "pinned" to the value $+1$. As a result, the spin–spin correlation $\langle s_1 s_2 \rangle$ is simply $\langle s_2 \rangle$. In Section 2.2, we saw that $\langle s_1 s_2 \rangle$ is equal to $\tanh K$. Therefore, in this limit,

$$\langle s_1 s_2 \rangle = \tanh K = \langle s_2 \rangle.$$

Now, we also saw in Chapter 2 (Equation 2.8) that the average value of a single spin $\langle s_i \rangle$ acted on by a magnetic field h is given by $\langle s_i \rangle = \tanh h$. In our case, the spin s_2 is acted on by h', and hence its average value is

$$\langle s_2 \rangle = \tanh h'.$$

Comparing the above two equations, we see that $h' = K$.

This result is also obtained from the transformation equations. We see from Equations 3.5 and 3.6 that the partial partition functions in this limit are

$$Z_+ \to e^{K+h}$$
$$Z_- \to e^{-K+h}.$$

Substituting into Equation 3.10 yields

$$h' = \frac{1}{2} \ln \left(\frac{Z_+}{Z_-} \right) \to \frac{1}{2} \ln \left(\frac{e^{K+h}}{e^{-K+h}} \right) = K.$$

This verifies the limit indicated by the dashed horizontal line in Figure 3.2.

Consistency Check

This is a good point to stop for a moment and check the internal consistency of the partial-summation transformation. We've said that the effective magnetic field h' in the primed system produces the same physics as the interactions K and h in the original system. Let's put this statement to the test with an explicit calculation.

First, we've pointed out in our previous discussion that from Equation 2.8 we can write the average value of the spin s_2 as

$$\langle s_2 \rangle = \tanh h'.$$

This follows because, *in the primed system*, we have a single spin s_2 acted on by the magnetic field h'.

On the other hand, we can calculate the average of s_2 by doing a sum over probabilities *in the original system*. Following this approach, we find

$$\langle s_2 \rangle = \frac{1}{Z} \sum_{\{s_1, s_2\}} s_2 e^{-\beta H} = \frac{Z_+ - Z_-}{Z_+ + Z_-} = \frac{e^{K+h} + e^{-K-h} - e^{-K+h} - e^{K-h}}{e^{K+h} + e^{-K-h} + e^{-K+h} + e^{K-h}}.$$

The denominator in this expression is simply the total partition function Z; namely, it's the sum of the partial partition functions for s_2 equal to $+1$ (Z_+) and for s_2 equal to -1 (Z_-). The numerator is $+1$ times the partial partition function where $s_2 = +1$ (Z_+), and (-1) times the partial partition function where $s_2 = -1$, (Z_-); that is, $Z_+ - Z_-$.

Now, at first glance these two expressions for $\langle s_2 \rangle$ look quite different. However, if we refer back to Equation 3.8, we can write the partial partition functions in terms of the primed quantities as follows:

$$\langle s_2 \rangle = \frac{Z_+ - Z_-}{Z_+ + Z_-} = \frac{e^{h' + K_0'} - e^{-h' + K_0'}}{e^{h' + K_0'} + e^{-h' + K_0'}}.$$

Notice that $e^{K_0'}$ factors out of each term in this expression, leaving

$$\langle s_2 \rangle = \frac{e^{h'} - e^{-h'}}{e^{h'} + e^{-h'}} = \tanh h'.$$

This agrees with our previous result. We knew this check had to work, but it's satisfying to go through the calculation just to verify all the details.

Extending the Transformation

We can extend the partial-summation transformation of this section by simply considering a longer chain of spins. For example, in Figure 3.3 we show three spins in a row, with a magnetic field h acting only on the spin on the left end, s_1. This spin affects s_2, producing an effective field h' on it that is smaller than h, and s_2 in turn affects spin s_3, producing an effective field h'' on it that is smaller than h'. It's a bit like passing a bag of popcorn from one person to another at the movies, with each person taking a handful, leaving precious little for the person at the end of the line.

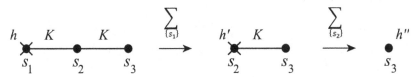

Figure 3.3 A three-spin system with a magnetic field h acting only on spin 1. Two partial summations generate an effective magnetic field h'' on spin 3.

To find the effective field acting on s_3, first do a partial sum on s_1, and then a partial sum on s_2. The first transformation, given in Equation 3.10, yields

$$h' = \frac{1}{2}\ln\left(\frac{e^{K+h} + e^{-K-h}}{e^{K-h} + e^{-K+h}}\right).$$

The next partial summation is just like the first, but with h' replacing h. Thus, we find

$$h'' = \frac{1}{2}\ln\left(\frac{e^{K+h'} + e^{-K-h'}}{e^{K-h'} + e^{-K+h'}}\right).$$

Clearly, this process can be extended to as long a chain of spins as desired.

We can take this calculation one step further, and express h'' solely in terms of the original K and h. Referring back to Equation 3.10, we can replace terms involving $e^{h'}$ with the following:

$$e^{h'} = \left(\frac{Z_+}{Z_-}\right)^{1/2}.$$

This gives

$$h'' = \frac{1}{2}\ln\left(\frac{e^K\left(\frac{Z_+}{Z_-}\right)^{1/2} + e^{-K}\left(\frac{Z_+}{Z_-}\right)^{-1/2}}{e^K\left(\frac{Z_+}{Z_-}\right)^{-1/2} + e^{-K}\left(\frac{Z_+}{Z_-}\right)^{1/2}}\right).$$

Simplifying, and writing out Z_+ and Z_- in terms of K and h, we find

$$h'' = \frac{1}{2}\ln\left(\frac{e^{2K+h} + 2e^{-h} + e^{-2K+h}}{e^{2K-h} + 2e^{h} + e^{-2K-h}}\right). \tag{3.12}$$

In a typical situation, it may be more convenient to calculate the numerical value of h' and h'', as opposed to writing out explicit expressions.

Cobwebbing

A convenient way to visualize the repeated transformations just described, and one that is quite useful in renormalization-group and mean-field theory calculations, is referred to as *cobwebbing*. An example is given in Figure 3.4, where the dashed lines that zigzag back and forth between the curve $y = h'$ and the straight-line $y = h$ are thought of as a cobweb filling in that space.

To see exactly how cobwebbing works, imagine a three-spin system with $K = 1$, and with a magnetic field $h = 2$ acting on s_1. This is indicated in the lower

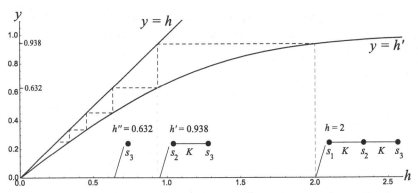

Figure 3.4 Cobwebbing to determine the values of h' and h'' for a three-spin system with $h = 2$ and $K = 1$.

right of Figure 3.4. To find the corresponding value of h', when s_1 is summed over, we go vertically to the curve $y = h'$. In this case, $h' = 0.938$, to three significant figures.

The next step is to sum over s_2 with a magnetic field equal to 0.938. To find the new effective field, h'', we start at the location 0.938 on the horizontal axis and go vertically until we hit the $y = h'$ curve. This yields the value $h'' = 0.632$. This is indicated in the lower middle of Figure 3.4.

A convenient graphical way to achieve the same result is to start at the point $h = 2$, $y = 0.938$, and move horizontally (constant y) to the left until you reach the $y = h$ line, as indicated by the top dashed line. At this point the values of y and h are equal, thus $h = 0.938$, just as desired for the next step of the process. You can now drop down vertically (constant h) to the curve $y = h'$, as indicated by the top vertical dashed line. This yields the new value of the effective field; namely, $h'' = 0.632$. This process of "cobwebbing" back and forth between $y = h'$ and $y = h$ can be continued indefinitely. In this case, we see that the effective field becomes smaller and smaller with each iteration of the process, asymptotically approaching $h = 0$.

3.3 Effective Two-Spin Interactions

To this point, we've considered systems where the partial-summation results in a single spin. In cases like these, the effective Hamiltonian consists only of zero-spin and one-spin terms. We now turn to systems where two spins remain after the partial summation, resulting in effective Hamiltonians with zero-, one-, and two-spin terms.

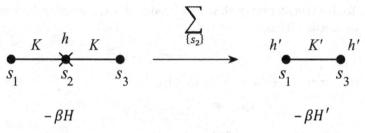

Figure 3.5 A partial-summation transformation from an original three-spin system to an effective two-spin system. Spin s_2 is summed over.

We start with the three-spin system shown on the left in Figure 3.5. This system has a magnetic field h acting on the middle spin s_2 – which will be summed out in the transformation – but no field acts on spins s_1 and s_3. (We'll consider a magnetic field on spins s_1 and s_3 later.) Even so, the nearest-neighbor coupling K between spin s_2 and spins s_1 and s_3 produces an effective magnetic field h' on s_1 and s_3, as indicated on the right side of Figure 3.5. In addition, the spins s_1 and s_3 experience an effective nearest-neighbor interaction K', even though they are second-neighbor spins with no direct interaction in the original system.

The Hamiltonian for the original system is

$$-\beta H = K(s_1 s_2 + s_2 s_3) + h s_2.$$

A partial sum over s_2 generates the following effective Hamiltonian for the two spins s_1 and s_3:

$$-\beta H' = 2K_0' + h'(s_1 + s_3) + K' s_1 s_3.$$

As before, we've included a zero-spin term K_0'. In fact, we've written it as $2K_0'$ because we want K_0' to represent the zero-spin term *per spin*, just like f is defined as the reduced free energy *per spin*. The Hamiltonian also includes an effective magnetic field h' acting on the spins s_1 and s_3, and an effective nearest-neighbor coupling K' between the spins.

We perform the transformation by considering each of the four spin configurations of s_1 and s_3 one at a time. Let's start with $s_1 = +1$ and $s_3 = +1$. Summing Boltzmann factors for the original system (i.e., summing over $s_2 = +1$ and $s_2 = -1$) yields the partial partition function we will call Z_{++}:

$$Z_{++} = \sum_{\{s_2, s_1=+1,\, s_3=+1\}} e^{-\beta H(\{s_2, s_1=+1, s_3=+1\})} = e^{2K+h} + e^{-2K-h}. \qquad (3.13)$$

The effective Hamiltonian, with $s_1 = +1$ and $s_3 = +1$, is equal to this partial partition function. Thus,

$$e^{2K_0' + 2h' + K'} = Z_{++} = e^{2K+h} + e^{-2K-h}.$$

Similarly, for $s_1 = +1$ and $s_3 = -1$ we have

$$e^{2K_0' + 0h' - K'} = Z_{+-} = e^h + e^{-h}.$$

For $s_1 = -1$ and $s_3 = +1$ we find

$$e^{2K_0' + 0h' - K'} = Z_{-+} = e^h + e^{-h}.$$

Notice that Z_{+-} and Z_{-+} are equal, as one would expect from the symmetry of the original system. We'll keep both symbols for now, however, just to make the structure of the transformation more transparent.

Finally, consider $s_1 = -1$ and $s_3 = -1$. In this case we have

$$e^{2K_0' - 2h' + K'} = Z_{--} = e^{-2K+h} + e^{2K-h}.$$

We see that Z_{--} is the same as Z_{++}, but with the signs of h and h' reversed. This is to be expected because reversing the sign of the spins reverses the sign of the one-spin terms, which have odd powers of the spins, but not the zero-spin or two-spin terms, which have even powers of the spins.

Let's take the log of both sides of these four equations, and collect them together for convenience. The result is

$$\begin{aligned}
2K_0' + 2h' + K' &= \ln Z_{++} = \ln(e^{2K+h} + e^{-2K-h}) \\
2K_0' + 0h' - K' &= \ln Z_{+-} = \ln(e^h + e^{-h}) \\
2K_0' + 0h' - K' &= \ln Z_{-+} = \ln(e^h + e^{-h}) \\
2K_0' - 2h' + K' &= \ln Z_{--} = \ln(e^{-2K+h} + e^{2K-h}).
\end{aligned} \tag{3.14}$$

Recall that the middle two equations are actually identical, and hence we have just three independent partial partition functions, Z_{++}, Z_{+-}, and Z_{--}, which can be used to determine the three independent Hamiltonian terms, K_0', h', and K'.

Solving a complex system of equations like this might seem a bit daunting at first, but it's actually quite simple. For any given term, just add together equations where that term is positive, and subtract equations where that term is negative. The simplest example of this approach is for h', which is positive in the first equation $(++)$ and negative in the last equation $(--)$. Hence, to solve for h' we subtract the last equation from the first equation. This yields

$$4h' = \ln Z_{++} - \ln Z_{--}$$
$$h' = \frac{1}{4}\ln\left(\frac{Z_{++}}{Z_{--}}\right) = \frac{1}{4}\ln\left(\frac{e^{2K+h} + e^{-2K-h}}{e^{-2K+h} + e^{2K-h}}\right). \tag{3.15}$$

Notice that h' depends on the *ratio* of partition functions.

The next simplest case is the zero-spin term K_0', which is positive in all four equations. Thus, we add these equations to find

$$8K_0' = \ln Z_{++} + \ln Z_{+-} + \ln Z_{-+} + \ln Z_{--}$$
$$K_0' = \frac{1}{8}\ln(Z_{++}\, Z_{+-}\, Z_{-+}\, Z_{--})$$
$$= \frac{1}{8}\ln\left[(e^{2K+h} + e^{-2K-h})(e^h + e^{-h})^2(e^{-2K+h} + e^{2K-h})\right]. \tag{3.16}$$

This result shows that K_0' is related to the *product* of all the partial partition functions.

Finally, to obtain K' we start with the first equation $(++)$, subtract the second equation $(+-)$, subtract the third equation $(-+)$, and then add the fourth equation $(--)$. This gives

$$4K' = \ln Z_{++} - \ln Z_{+-} - \ln Z_{-+} + \ln Z_{--}$$
$$K' = \frac{1}{4}\ln\left(\frac{Z_{++}\, Z_{--}}{Z_{+-}\, Z_{-+}}\right). \tag{3.17}$$

Thus, the effective two-spin coupling depends on the product of two partial partition functions *divided* by the product of two other partial partition functions. Substituting the appropriate partial partition functions yields K' as a function of the original K and h:

$$K' = \frac{1}{4}\ln\left[\frac{(e^{2K+h} + e^{-2K-h})(e^{-2K+h} + e^{2K-h})}{(e^h + e^{-h})^2}\right]. \tag{3.18}$$

In the simple case where $h = 0$, we find that $h' = 0$ – as expected, since the original system is now up–down symmetric. The nearest-neighbor coupling simplifies in this case to

$$K' = \frac{1}{2}\ln(\cosh 2K). \tag{3.19}$$

This result is shown by the curve in Figure 3.6, along with the dashed line showing $y = K$. Notice that K' is always less than K, as one would expect. In addition, $K' \to 0$ in the limit $K \to 0$, and $K' \to K - \frac{1}{2}\ln 2$ in the limit $K \to \infty$.

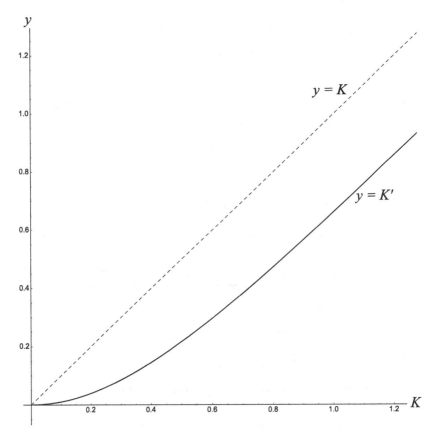

Figure 3.6 A plot of $y = K'$ (curve) as a function of K. The dashed line is $y = K$.

Extending the Transformation

A straightforward extension of the transformation shown in Figure 3.5 is to apply the transformation in two stages, as indicated in Figure 3.7. We start by reducing a five-spin system to a three-spin system in the first stage, and then we reduce the three-spin system to a two-spin system in the second stage. To keep things simple, we'll consider the case of zero magnetic field.

In the first stage, we sum over the spins s_2 and s_4, as indicated on the left side of Figure 3.7. These sums can be performed independently of one another because – and here's the key point – there are no terms in the original Hamiltonian that directly couple these spins. That is, no terms like Ls_2s_4, for example. As a result, each summation produces the transformation derived in Equation 3.19:

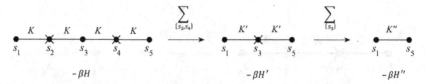

Figure 3.7 Applying the partial-summation transformation of Figure 3.5 in two stages reduces the original five-spin system with interaction K to an equivalent two-spin system with interaction K''.

$$K' = \frac{1}{2}\ln(\cosh 2K).$$

The zero-spin term in Equation 3.16 is also generated for each sum.

In the second stage, shown in the middle of Figure 3.7, we simply repeat the transformation described in Figure 3.5, this time with K' replacing K. Thus, summing over s_3 generates the following effective coupling between s_1 and s_5:

$$K'' = \frac{1}{2}\ln(\cosh 2K'). \tag{3.20}$$

This process can be extended to as large a lattice as desired by simply including more stages to the process. We'll explore this possibility in Chapter 7.

A Different Extension

So far, all our transformations have involved the summation of a single spin to generate an effective Hamiltonian. There is no limit to the number of spins we can sum, however. As an example, we can sum over the two spins indicated with an "X" in Figure 3.8 to generate an effective Hamiltonian between the remaining two spins.

The original Hamiltonian is

$$-\beta H = K(s_1 s_2 + s_2 s_3 + s_3 s_4).$$

Notice that we are setting the magnetic field equal to zero for this example. We now sum over the four configurations of the two interior spins, s_2 and s_3, to generate an effective Hamiltonian for the remaining spins, s_1 and s_4. This effective Hamiltonian is

$$-\beta H' = 2K'_0 + K' s_1 s_4.$$

Notice that we've included a zero-spin term for each of the two spins in the primed system.

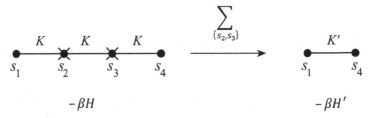

Figure 3.8 Summing two spins in a four-spin system to generate an effective Hamiltonian for the remaining two spins.

The transformation in Figure 3.8 has the same basic structure as the one in Figure 3.5, except now each of the partial partition functions has contributions from four spin configurations – as opposed to just two. In addition, there are only two independent partial partition functions now. One is Z_{++}, which is equal to Z_{--} because the magnetic field is zero; the other is Z_{+-}, which is equal to Z_{-+} because of the left–right symmetry of the system (and also because of the up–down symmetry). It follows that the renormalized coupling, K' – in terms of the partial partition functions – is given by Equation 3.17, which reduces to the following:

$$K' = \frac{1}{4} \ln \left(\frac{Z_{++} \, Z_{--}}{Z_{+-} \, Z_{-+}} \right) = \frac{1}{2} \ln \left(\frac{Z_{++}}{Z_{+-}} \right).$$

All that remains is to calculate Z_{++} and Z_{+-} for this new system.

We'll start with Z_{++}, which corresponds to $s_1 = +1$ and $s_4 = +1$, and follow that with Z_{+-}, which corresponds to $s_1 = +1$ and $s_4 = -1$. The results are

$$2K_0' + K' = \ln Z_{++} = \ln(e^{3K} \quad + \quad e^{-K} \quad + \quad e^{-K} \quad + \quad e^{-K})$$
$$\phantom{2K_0' + K' = \ln Z_{++} = \ln(} {+}{+}{+}{+} \quad\quad {+}{+}{-}{+} \quad\quad {+}{-}{+}{+} \quad\quad {+}{-}{-}{+}$$

$$2K_0' - K' = \ln Z_{+-} = \ln(e^{K} \quad + \quad e^{K} \quad + \quad e^{-3K} \quad + \quad e^{K})$$
$$\phantom{2K_0' - K' = \ln Z_{+-} = \ln(} {+}{+}{+}{-} \quad\quad {+}{+}{-}{-} \quad\quad {+}{-}{+}{-} \quad\quad {+}{-}{-}{-}. \quad (3.21)$$

Below each term we show the spin assignments, for s_1 through s_4. Subtract the second equation from the first equation to find

$$K' = \frac{1}{2} \ln \left(\frac{e^{3K} + 3e^{-K}}{e^{-3K} + 3e^{K}} \right). \tag{3.22}$$

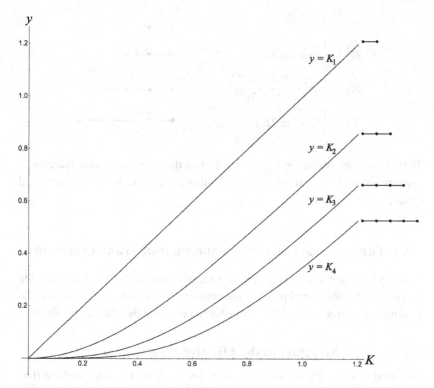

Figure 3.9 Interactions between the two end spins of a group for increasing separation. The subscripts indicate the number of bonds connecting the end spins.

Sum the two equations to find the zero-spin term:

$$K'_0 = \frac{1}{8}\ln(Z_{++}\, Z_{+-}\, Z_{-+}\, Z_{--}) = \frac{1}{4}\ln(Z_{++}\, Z_{+-})$$
$$= \frac{1}{4}\ln[(e^{3K} + 3e^{-K})(e^{-3K} + 3e^{K})]. \tag{3.23}$$

Let's collect the various results we've obtained for the effective two-spin interaction and plot them together for comparison. To keep things straight, we'll number our results K_1 through K_4 as follows: K_1 is the original two-spin interaction with one bond connecting the spins on either end; K_2 is the effective two-spin interaction with two bonds connecting the end spins (Equation 3.19); K_3 is the effective two-spin interaction with three bonds connecting the end spins (Equation 3.22); and K_4 is the effective two-spin interaction with four bonds connecting the end spins (Equation 3.20). Here are the corresponding relations:

$$K_1 = K$$

$$K_2 = \frac{1}{2}\ln(\cosh 2K)$$

$$K_3 = \frac{1}{2}\ln\left(\frac{e^{3K} + 3e^{-K}}{e^{-3K} + 3e^{K}}\right)$$

$$K_4 = \frac{1}{2}\ln(\cosh 2K_2)$$

These results are plotted in Figure 3.9. Notice that each successive increase in distance between the end spins gives a smaller effective coupling, as one would expect.

3.4 The Structure of Partial-Summation Transformations

Looking back at the partial-summation transformations we've derived in the previous sections, several patterns are evident. In this section, we examine the fundamental "structure" of these transformations, and the reasons for them.

Structure of the Effective Hamiltonian

The most obvious feature of partial-summation transformations is the difference in the way the zero-spin terms transform as compared with the one-spin and two-spin terms. Recalling the first transformation we derived, which was in Section 3.2 and was for the case of summing one spin and leaving one spin unsummed, we found the following:

$$K_0' = \frac{1}{2}\ln(Z_+ Z_-)$$

$$h' = \frac{1}{2}\ln\left(\frac{Z_+}{Z_-}\right). \tag{3.24}$$

Notice that K_0' depends on the log of products of the partial partition functions, while h' depends on the log of the ratio of partial partition functions.

We saw this again in Section 3.3, when we summed one spin and left two spins unsummed. In that case we found

$$K_0' = \frac{1}{8}\ln(Z_{++} \ Z_{+-} \ Z_{-+} \ Z_{--})$$

$$h' = \frac{1}{4}\ln\left(\frac{Z_{++}}{Z_{--}}\right)$$

$$K' = \frac{1}{4}\ln\left(\frac{Z_{++} \ Z_{--}}{Z_{+-} \ Z_{-+}}\right). \tag{3.25}$$

Again, K_0' depends on the log of the product of the partial partition functions, and the interactions (h' and K') depend on the log of ratios of partial partition functions. Note in addition that the interactions have an equal number of Zs in both the numerator and the denominator.

The reason for this structure in the transformation can be understood as follows. First, K_0' has a product-type dependence because the log of a product is equal to the sum of the logs – thus K_0' is a sum over all of the contributions to the free energy from the various partial partition functions. Next, the reason that interactions have an equal number of Zs in the numerator and denominator of the log becomes clear when one considers zero coupling. In this limit, all the Zs are equal to one another, and hence all the renormalized couplings are proportional to the log of 1. As a result, the renormalized couplings are equal to zero as well, as one would expect.

Adding a Magnetic Field to the Unsummed Spins

Near the beginning of Section 3.3, we considered a transformation where there was a magnetic field acting on the spin that was summed over, but no magnetic fields on the remaining two spins. We mentioned at the time that we would add a magnetic field on the end spins later, so let's do that now.

Adding a magnetic field h to the end spins introduces an overall factor of e^{2h} to each term of Z_{++} – that is, e^h for each of the two plus spins. It also multiplies each term of Z_{+-} by e^{0h}, multiplies each term of Z_{-+} by e^{0h}, and multiplies each term of Z_{--} by e^{-2h}. Let's see how these changes affect the results of the transformation.

First, consider h':

$$h' = \frac{1}{4}\ln\left(\frac{Z_{++}}{Z_{--}}\right) \rightarrow \frac{1}{4}\ln\left(\frac{e^{2h}Z_{++}}{e^{-2h}Z_{--}}\right) = h + \frac{1}{4}\ln\left(\frac{Z_{++}}{Z_{--}}\right). \tag{3.26}$$

Thus, the added magnetic field h just adds directly to the effective magnetic field already calculated.

Next, what about the two-spin term? In this case we find

$$K' = \frac{1}{4}\ln\left(\frac{Z_{++}\,Z_{--}}{Z_{+-}\,Z_{-+}}\right) \rightarrow \frac{1}{4}\ln\left(\frac{e^{2h}Z_{++}\,e^{-2h}Z_{--}}{e^{0h}Z_{+-}\,e^{0h}Z_{-+}}\right) = \frac{1}{4}\ln\left(\frac{Z_{++}\,Z_{--}}{Z_{+-}\,Z_{-+}}\right).$$

Notice that the added magnetic field cancels out and produces no change to the two-spin interaction. Similarly, for the zero-spin term we find

$$K_0' = \frac{1}{8}\ln(Z_{++} \ Z_{+-} \ Z_{-+} \ Z_{--}) \rightarrow \frac{1}{8}\ln(e^{2h}Z_{++} \ e^{0h}Z_{+-} \ e^{0h}Z_{-+} \ e^{-2h}Z_{--}).$$

Once again, the additional factors due to the magnetic field cancel, leaving the same zero-spin term as before. Thus, the only effect of the added magnetic field is that it adds directly to the renormalized magnetic field.

3.5 Transforming the Free Energy

We've mentioned several times that K_0' contributes to the free energy of a system. In this section we explore the connection in detail.

To begin, the basic requirement in all our transformations is that the partition function be preserved. That is,

$$Z = Z'.$$

Recall that these two partition functions are defined as follows:

$$Z = \sum_{\{\mu_i\}} \sum_{\{\sigma_i\}} e^{-\beta H(\{\mu_i\},\{\sigma_i\})}$$

$$Z' = \sum_{\{\mu_i\}} e^{-\beta H'(\{\mu_i\})}.$$

In these expressions, σ_i represents a group of spins that are summed over, and μ_i represents spins that remain.

In all of our transformations, the reduced primed Hamiltonian, $-\beta H'(\{\mu_i\})$, contains the term $N'K_0'$, where N' is the number of spins that remain after summing. We can factor this term out, leaving a Hamiltonian that depends only on the *interactions* in the primed system. Referring to this reduced Hamiltonian as $-\beta H'_{\text{int}}(\{\mu_i\})$, we can write

$$Z' = \sum_{\{\mu_i\}} e^{N'K_0'}e^{-\beta H'_{\text{int}}(\{\mu_i\})} = e^{N'K_0'}\sum_{\{\mu_i\}} e^{-\beta H'_{\text{int}}(\{\mu_i\})} = e^{N'K_0'}Z'_{\text{int}}. \qquad (3.27)$$

The final term in this expression, Z'_{int}, is the partition function of the primed interaction Hamiltonian,

$$Z'_{\text{int}} = \sum_{\{\mu_i\}} e^{-\beta H'_{\text{int}}(\{\mu_i\})}. \qquad (3.28)$$

The corresponding reduced free energy is

$$f'_{int} = \frac{1}{N'} \ln Z'_{int}.$$
(3.29)

Now, let's take a look at the reduced free energy of the original system, f, which is the ultimate goal of our calculation. We have the following, based on preservation of the partition function:

$$f = \frac{1}{N} \ln Z = \frac{1}{N} \ln Z' = \frac{1}{N} \ln(e^{N'K'_0} Z'_{int}).$$

Inserting N'/N' and expanding the log, we have

$$f = \frac{N'}{N} \frac{1}{N'} \ln(e^{N'K'_0} Z'_{int}) = \frac{N'}{N} \left(\frac{1}{N'} N'K'_0 + \frac{1}{N'} \ln Z'_{int} \right).$$
(3.30)

Finally, simplifying and using the definition of f'_{int}, we find

$$f = \frac{N'}{N} (K'_0 + f'_{int}).$$
(3.31)

Here we see that the reduced free energy we desire depends not only on the reduced free energy of the interactions in the primed system, but also on the contributions contained in K'_0.

A Specific Example

Let's apply the free energy relationship in Equation 3.31 to a specific case; namely, two spins in the original system and one spin in the primed system, as in Figure 3.1. Therefore, $N = 2$ and $N' = 1$ and

$$f = \frac{1}{2} (K'_0 + f'_{int}).$$
(3.32)

The original reduced free energy is

$$f = \frac{1}{2} \ln Z = \frac{1}{2} \ln(Z_+ + Z_-).$$
(3.33)

Expressing the result in terms of the partial partition functions Z_+ and Z_- is helpful in simplifying the algebraic steps to come.

Next, let's evaluate the terms on the right-hand side of Equation 3.32. First, referring back to Section 3.2, we have

$$K'_0 = \frac{1}{2} \ln(Z_+ Z_-) = \ln(Z_+^{1/2} Z_-^{1/2}).$$
(3.34)

In addition, the primed system is simply a single spin acted on by the magnetic field, h', where

$$h' = \frac{1}{2}\ln\left(\frac{Z_+}{Z_-}\right). \tag{3.35}$$

The corresponding reduced free energy for this interaction is

$$f'_{\text{int}} = \ln(e^{h'} + e^{-h'}) = \ln\left(\left(\frac{Z_+}{Z_-}\right)^{1/2} + \left(\frac{Z_-}{Z_+}\right)^{1/2}\right) = \ln\left(\frac{Z_+ + Z_-}{Z_+^{1/2} Z_-^{1/2}}\right). \tag{3.36}$$

Combining Equations 3.34 and 3.36, it follows that

$$K'_0 + f'_{\text{int}} = \ln(Z_+ + Z_-).$$

This validates the free energy relationships in Equations 3.32 and 3.33.

3.6 The Star-Triangle Transformation

A partial-summation transformation that plays an important role in the study of Ising models on triangular and hexagonal lattices is the *star-triangle transformation*. The origin of the name of this transformation is apparent in Figure 3.10, which shows the original "star" system on the left, and the primed "triangle" system on the right. To carry out this transformation, the center spin in the star, s_4, is summed over, leaving an effective Hamiltonian for the remaining spins in the triangle.

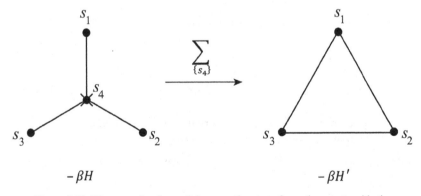

Figure 3.10 The star-triangle partial-summation transformation starts with the star-shaped system on the left, and, after summing over the center spin, s_4, generates the triangle-shaped system on the right.

The original Hamiltonian is

$$-\beta H = K(s_1 + s_2 + s_3)s_4 + hs_4. \tag{3.37}$$

A simpler Hamiltonian would omit the magnetic field on s_4, but we include it to show an important new feature of partial-summation transformations.

This new feature is apparent in the effective Hamiltonian for the triangle:

$$-\beta H' = 3K'_0 + h'(s_1 + s_2 + s_3) + K'(s_1s_2 + s_2s_3 + s_3s_1) + P's_1s_2s_3. \tag{3.38}$$

We see familiar terms in this Hamiltonian, like a zero-spin term for each of the three spins in the primed system, $3K'_0$, an effective magnetic field on all the spins, h', and an effective nearest-neighbor coupling acting on all pairs of spins, K'. What's surprising – and totally unexpected based on what we've seen so far – is the new term P' that couples all three spins. Where did this three-spin interaction come from (who ordered that?), and what role does it play in the system?

To answer these questions, we start by looking at the number of independent partial partition functions in the system, since this will determine the number of independent terms in the effective Hamiltonian. We'll label the partial partition functions with subscripts that indicate the sign of the spins s_1, s_2, and s_3. For example, Z_{+++} refers to the partial partition function with $s_1 = +1$, $s_2 = +1$, and $s_3 = +1$. Similarly, Z_{++-} refers to the case $s_1 = +1$, $s_2 = +1$, and $s_3 = -1$.

Now, Z_{+++} and Z_{---} are different independent functions because the magnetic field on s_4 breaks the up–down symmetry. The functions Z_{++-} and Z_{--+} are independent for the same reason. This gives four independent functions for the system, implying four independent terms in the effective Hamiltonian. There are no other independent partial partition functions because $Z_{++-} = Z_{+-+} = Z_{-++}$ and $Z_{--+} = Z_{-+-} = Z_{+--}$ due to the triangular symmetry of the system.

When we consider possible Hamiltonians for the triangle in Figure 3.10, we start with zero-spin, one-spin, and two-spin terms. These represent only three different terms, however, and we need four. The *only* other possibility is a three-spin interaction, and hence the inclusion of the term $P's_1s_2s_3$.

The three-spin term wouldn't appear in the absence of a magnetic field. This is because h breaks the up–down symmetry, which allows for other terms like $P's_1s_2s_3$ that also lack this symmetry. If s_4 connected to five different spins, and $h \neq 0$, then we would expect five-spin terms to appear in the Hamiltonian as well.

Let's carry out the transformation. Considering each of the four independent choices for s_1, s_2, and s_3, we have the following four independent equations:

$$
\begin{aligned}
3K_0' + 3h' + 3K' + P' = \ln Z_{+++} &= \ln(e^{3K+h} + e^{-3K-h}) \\
3K_0' + h' - K' - P' = \ln Z_{++-} &= \ln(e^{K+h} + e^{-K-h}) \\
3K_0' - h' - K' + P' = \ln Z_{--+} &= \ln(e^{-K+h} + e^{K-h}) \\
3K_0' - 3h' + 3K' - P' = \ln Z_{---} &= \ln(e^{-3K+h} + e^{3K-h}).
\end{aligned} \quad (3.39)
$$

To solve for each term, we add equations where that term is positive and subtract equations where that term is negative, always keeping in mind the possibility that we may need to multiply one or more of the equations by an overall factor to make things work out properly. The results are as follows:

$$
\begin{aligned}
K_0' &= \frac{1}{24} \ln(Z_{+++} \, Z_{++-}{}^3 \, Z_{--+}{}^3 \, Z_{---}) \\
h' &= \frac{1}{8} \ln\left(\frac{Z_{+++} \, Z_{++-}}{Z_{---} \, Z_{--+}}\right) \\
K' &= \frac{1}{8} \ln\left(\frac{Z_{+++} \, Z_{---}}{Z_{++-} \, Z_{--+}}\right) \\
P' &= \frac{1}{8} \ln\left(\frac{Z_{+++} \, Z_{--+}{}^3}{Z_{---} \, Z_{++-}{}^3}\right).
\end{aligned} \quad (3.40)
$$

The star-triangle transformation is crucial in connecting the critical properties of triangular and hexagonal lattices.

At this point, one might wonder why there are no other new interactions in the primed Hamiltonian. After all, we found the need for a new three-spin interaction, $P' s_1 s_2 s_3$, so perhaps other interactions may be needed as well. For example, consider a four-spin interaction like $Q' s_1 s_1 s_2 s_3$. One might consider adding this term to the Hamiltonian, but when you realize that $s_1^2 = 1$, it follows that this interaction reduces to an interaction we already have, $K' s_2 s_3$. Similar reasoning eliminates any other interaction that might be proposed.

3.7 Problems

3.1 Refer to Figure 3.1 for the following questions. (a) Is Z_+ greater than, less than, or equal to Z_- in the zero-coupling limit? Explain. (b) What are Z_+ and Z_- in the zero-coupling limit?

3.2 Consider the partial-summation transformation shown in Figure 3.11, where the configurations of spin s_1 are summed over to generate an effective Hamiltonian for spin s_2.

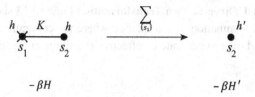

Figure 3.11 A partial-summation transformation from two spins to one spin.

The original (unprimed) and effective (primed) Hamiltonians for this system are as follows:

$$-\beta H = Ks_1s_2 + h(s_1 + s_2)$$
$$-\beta H' = K'_0 + h's_2.$$

(a) What are Z_+ and Z_- for this system? (b) Find K'_0 and compare your answer to the result given in Equation 3.11. (c) Find h' and compare your answer to the result given in Equation 3.10.

3.3 Consider the partial-summation transformation shown in Figure 3.12, where the configurations of spins s_1 and s_2 are summed over to generate an effective Hamiltonian for the remaining spin, s_3.

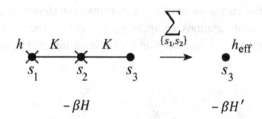

Figure 3.12 A partial-summation transformation from three spins to one spin.

The original (unprimed) and effective (primed) Hamiltonians for this system are as follows:

$$-\beta H = K(s_1s_2 + s_2s_3) + hs_1$$
$$-\beta H' = K'_0 + h_{\text{eff}}s_3.$$

(a) Do you expect h_{eff} to be greater than, less than, or equal to h'' given in Equation 3.12? Explain. (b) Find h_{eff} and compare with Equation 3.12. (c) Find K'_0 for this transformation.

3.4 **Generalized Three-to-Two Transformation** Figure 3.13 shows a generalized partial-summation transformation where the configurations of spin s_2 are summed over to generate an effective Hamiltonian for spins s_1 and s_3.

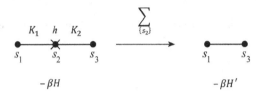

Figure 3.13 A generalized partial-summation transformation from three spins to two spins.

The Hamiltonian for the original system is

$$-\beta H = K_1 s_1 s_2 + K_2 s_2 s_3 + h s_2.$$

(a) How many independent partial partition functions are there for this transformation? (b) Write out the effective Hamiltonian $-\beta H'$ in terms of primed quantities. (c) Determine each of the primed quantities in $-\beta H'$ in terms of K_1, K_2, and h.

3.5 Consider the partial-summation transformation shown in Figure 3.14, where the configurations of spins s_2, s_3, and s_4 are summed over to generate an effective Hamiltonian for the remaining spins, s_1 and s_5.

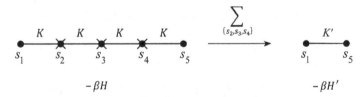

Figure 3.14 A partial-summation transformation from five spins to two spins.

The original (unprimed) and effective (primed) Hamiltonians for this system are as follows:

$$-\beta H = K(s_1 s_2 + s_2 s_3 + s_3 s_4 + s_4 s_5)$$
$$-\beta H' = 2K_0' + K' s_1 s_5.$$

(a) Which of the following expressions is the partial partition function Z_{++} for this system?

$$Z_1 = e^{4K} + 4 + e^{-4K}$$
$$Z_2 = e^{3K} + 6 + e^{-3K}$$
$$Z_3 = e^{4K} + 6 + e^{-4K}$$
$$Z_4 = e^{3K} + 8 + e^{-3K}.$$

(b) For each of the other expressions, give at least one reason why it cannot be Z_{++}. (c) What is K'_0 in the zero-coupling limit?

3.6 **The Generalized Star-Triangle Transformation** Figure 3.15 shows a star-triangle transformation in which the strength of the coupling to spins s_1, s_2, and s_3 in the original system can be different.

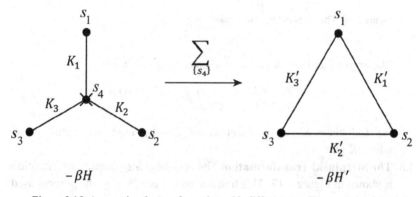

Figure 3.15 A star-triangle transformation with different coupling to each of the outer spins of the original system.

The original (unprimed) and effective (primed) Hamiltonians for this system are as follows:

$$-\beta H = K_1 s_1 s_4 + K_2 s_2 s_4 + K_3 s_3 s_4$$
$$-\beta H' = 3K'_0 + K'_1 s_1 s_2 + K'_2 s_2 s_3 + K'_3 s_3 s_1.$$

(a) If $K_1 = 0$ and $K_2 = K_3 > 0$, is K'_1 greater than, less than, or equal to K'_2? Explain. (b) Assuming K_1, K_2, and K_3 all have different values, how many independent partial partition functions are there in this system? (c) Derive the expressions for K'_1, K'_2, and K'_3. (d) Show that $K_1 = 0$, $K_2 = K_3 = K$ yields the expected results for K'_1, K'_2, and K'_3.

3.7 **The Diamond Transformation** The diamond transformation is illustrated in Figure 3.16. In this case, two of the four original spins are summed over to generate an effective Hamiltonian for the remaining two spins.

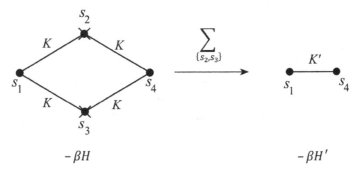

Figure 3.16 The diamond transformation.

The original and effective Hamiltonians are given below:

$$-\beta H = K(s_1 s_2 + s_1 s_3 + s_2 s_4 + s_3 s_4)$$
$$-\beta H' = 2K'_0 + K' s_1 s_4.$$

(a) Calculate K' as a function of K. (b) Find the value of K where $K'(K) = K$.

3.8 **The Sierpinski Transformation** The zero-field Sierpinski transformation is shown in Figure 3.17. This transformation is related to the process used to generate the fractal known as the Sierpinski Gasket.

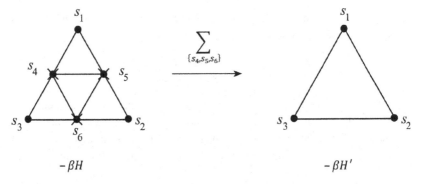

Figure 3.17 The Sierpinski transformation.

The original and effective Hamiltonians are given below:

$$-\beta H = K(s_1 s_4 + s_1 s_5 + s_4 s_5 + s_4 s_6 + s_5 s_6 + s_3 s_6 + s_6 s_2 + s_3 s_4 + s_2 s_5)$$
$$-\beta H' = 3K_0' + K'(s_1 s_2 + s_2 s_3 + s_3 s_1).$$

(a) Calculate Z_{+++}. (b) Calculate Z_{++-}.

4

Infinite Ising Systems in One Dimension

In the last chapter we explored transformations where a finite group of Ising spins is summed to produce effective interactions among the remaining spins. In all of these cases, a finite sum of Boltzmann factors is sufficient to solve the problem. We turn now to infinite systems, where a straightforward, brute-force summation is not possible. Instead, we develop several techniques that allow us to evaluate an infinite summation in full detail.

4.1 The Nibble Approach

One of the transformations studied in Chapter 3 involved summing the end spin in a chain of Ising spins to produce an effective Hamiltonian. In this section we show how this transformation can be applied to an infinite chain of spins. We refer to this as the *Nibble approach* because of the way the calculation "nibbles" away at the problem by summing (indicated with an X) one spin at a time along the chain – like a character chomping a row of dots in a video game, or a chicken gobbling up a row of seeds. Each partial summation is performed quite easily, and when the results of N summations are combined the system is solved. This method is an excellent way to solve one-dimensional Ising systems, but it cannot be used in higher dimensions for reasons we'll explore as we go along.

To see how this method works, we'll start with just two spins, as in Figure 4.1. The two-spin partition function Z_2 for this system is

$$Z_2 = \sum_{\{s_1\}} \sum_{\{s_2\}} e^{Ks_1 s_2}.$$

94

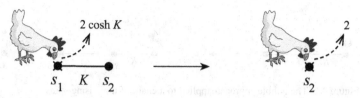

Figure 4.1 When spin s_1 is nibbled off, a factor $(2\cosh K)$ is generated. When spin s_2 is nibbled off, an additional factor of 2 is generated.

First, we'll nibble away at the system by summing over the two configurations of spin s_1. The result is

$$Z_2 = \sum_{\{s_2\}}(e^{Ks_2} + e^{-Ks_2}).$$

The fact that s_2 can only take on the values $+1$ and -1 leads to the following relationship:

$$(e^{Ks_2} + e^{-Ks_2}) = (2\cosh K).$$

This can be verified by direct substitution of the values of s_2 into the equation. With this result in hand, we can complete the evaluation of Z_2 as follows:

$$Z_2 = \sum_{\{s_2\}}(e^{Ks_2} + e^{-Ks_2}) = (2\cosh K)\sum_{\{s_2\}}1 = (2\cosh K)(2). \qquad (4.1)$$

Thus, summing over s_1 produces a factor of $(2\cosh K)$, and summing over s_2 produces a factor of 2.

To take the process one step further, consider three spins, as in Figure 4.2. In this case, the three-spin partition function Z_3 is

$$Z_3 = \sum_{\{s_1\}}\sum_{\{s_2\}}\sum_{\{s_3\}}e^{Ks_1s_2}e^{Ks_2s_3} = \sum_{\{s_2\}}\sum_{\{s_3\}}(e^{Ks_2} + e^{-Ks_2})e^{Ks_2s_3}$$

$$= (2\cosh K)\sum_{\{s_3\}}(e^{Ks_3} + e^{-Ks_3}) = (2\cosh K)(2\cosh K)(2)$$

$$= 2(2\cosh K)^2.$$

Thus, nibbling off a spin produces a factor of $(2\cosh K)$, until the last spin in the chain, when a factor of 2 is produced.

It follows that for N spins in a chain, where N is any positive integer, the partition function is

$$Z_N = 2(2\cosh K)^{N-1} = 2^N(\cosh K)^{N-1}. \qquad (4.2)$$

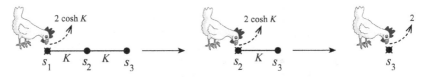

Figure 4.2 The Nibble approach applied to a chain of three Ising spins.

This is a significant result – it gives us the partition function for a chain of spins of arbitrary length N. Thus, rather than carrying out an explicit summation over 2^N configurations, which would be the brute-force approach, we can simply substitute N into Equation 4.2 to obtain the same result.

Free Energy

Using the result just obtained, the reduced free energy per site for an Ising chain of N sites is

$$f_N(K) = \frac{1}{N}\ln Z_N = \ln 2 + \frac{N-1}{N}\ln(\cosh K). \qquad (4.3)$$

Let's take a look at some of the key limits of this result. First, if $K \to 0$ we have $\cosh K \to 1$, which implies $\ln(\cosh K) \to 0$. As a result,

$$\lim_{K \to 0} f_N(K) = \ln 2.$$

This is the expected "all entropy" result for zero coupling.

In the limit $K \to \infty$ we have

$$\lim_{K \to \infty} f_N(K) = \ln 2 + \frac{N-1}{N}\ln\left(\frac{e^K}{2}\right) = \ln 2 + \left(\frac{N-1}{N}\right)(K - \ln 2)$$
$$= \left(\frac{N-1}{N}\right)K + \frac{1}{N}\ln 2.$$

The first term in this limit is the ground-state energy per site; namely, K multiplied by the number of bonds $(N-1)$ divided by the number of sites (N). The second term is $\ln 2$ divided by N, which gives the residual entropy *per site* of the two ground-state spin configurations – all up and all down.

In the thermodynamic limit $(N \to \infty)$ we find the following:

$$f(K) = \lim_{N \to \infty} f_N(K) = \lim_{N \to \infty}\left[\ln 2 + \frac{N-1}{N}\ln(\cosh K)\right].$$

Noting that $(N-1)/N$ approaches 1 as $N \to \infty$, we have

$$f(K) = \ln 2 + \ln(\cosh K). \tag{4.4}$$

This is a result that simply can't be obtained with a brute-force calculation.

In the limit of zero coupling, the reduced free energy per site for the infinite lattice is $\ln 2$, as expected:

$$\lim_{K \to 0} f(K) = \ln 2.$$

In the limit of infinite coupling, we have a much simpler result than was found in the finite case. Specifically,

$$\lim_{K \to \infty} f(K) = K.$$

Thus, we have one bond per site for the infinite lattice, giving a reduced energy K per site. In addition, there is no residual entropy *per site* from the two ground states of the infinite lattice.

Entropy and Specific Heat

Now that the reduced free energy per site has been obtained for both finite and infinite systems, we can use these results to calculate other thermodynamic functions of interest. For example, the reduced entropy per site for a finite system with N sites is

$$\begin{aligned} s_N &= f_N - K \frac{\partial f_N}{\partial K} \\ &= \ln 2 + \left(\frac{N-1}{N}\right)[\ln(\cosh K) - K \tanh K]. \end{aligned}$$

For an infinite system, the corresponding entropy is

$$\begin{aligned} s &= f - K \frac{\partial f}{\partial K} \\ &= \ln 2 + \ln(\cosh K) - K \tanh K. \end{aligned}$$

Results for various values of N are plotted in Figure 4.3. In each case, the entropy is equal to $\ln 2$ at $K = 0$, as expected for zero coupling. In the limit $K \to \infty$, the entropy decreases to zero in the infinite case, but remains finite for finite values of N.

Next, let's consider the reduced specific heat per site. For the finite Ising chain, with N sites and free boundary conditions, we find

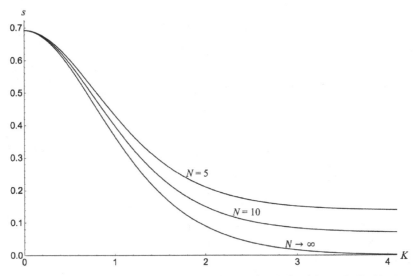

Figure 4.3 The reduced entropy s per site for an Ising chain with $N = 5$, $N = 10$, and $N \to \infty$.

$$c_N = K^2 \left(\frac{N - 1}{N}\right) \left(\frac{1}{\cosh K}\right)^2.$$

In the limit of an infinite system, we have

$$c = K^2 \left(\frac{1}{\cosh K}\right)^2. \tag{4.5}$$

These results are plotted in Figure 4.4 for various values of N. Notice that for each value of N there is a peak near $K = 1$. In addition, the specific heat goes to zero for both $K = 0$ and $K \to \infty$. Finally, we see that the finite-lattice results merge smoothly with the infinite-lattice results as N is increased.

Spin–Spin Correlations

The Nibble approach provides a useful way to study correlations between spins in a chain of Ising spins. The method is straightforward to carry out, and has an interesting graphical interpretation as well.

To begin, imagine a system of two spins with coupling K, as shown in Figure 4.1. As we've seen, the partition function for this system is

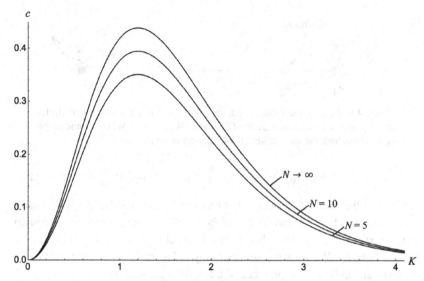

Figure 4.4 The reduced specific heat per site for an Ising chain with $N = 5$, $N = 10$, and $N \to \infty$. The specific heat reaches a peak for intermediate values of K, and goes to zero in the limit of large and small values of K.

$$Z_2 = \sum_{\{s_1\}} \sum_{\{s_2\}} e^{Ks_1s_2} = 2(2\cosh K).$$

Suppose we would like to find the average value of spin s_1 in this system. To do this we perform the following calculation:

$$\langle s_1 \rangle = \frac{1}{Z_2} \sum_{\{s_1\}} \sum_{\{s_2\}} s_1 e^{Ks_1s_2}.$$

We know this average value is zero, since there is no preference for a spin to be up or down, but it's instructive to see how this result is obtained with the Nibble approach.

We'll start by summing over s_1. The result is

$$\sum_{\{s_1\}} s_1 e^{Ks_1s_2} = (e^{Ks_2} - e^{-Ks_2}).$$

Now, it may seem surprising at first, but the quantity in parentheses is equal to $(2\sinh K)s_2$; that is, the spin s_2 starts out in the exponent, in the Boltzmann factors e^{Ks_2} and e^{-Ks_2}, but comes down to multiply the quantity $(2\sinh K)$. This can be verified by first substituting $s_2 = +1$, and then $s_2 = -1$, into the following:

Figure 4.5 When a site containing a spin is nibbled off, the spin (open circle) jumps to the next site, and a factor of $(2 \sinh K)$ is generated. When the end of the chain is reached, the summation (nibble) generates a factor of 0.

$$(e^{Ks_2} - e^{-Ks_2}) = (e^K - e^{-K})s_2 = (2\sinh K)s_2. \tag{4.6}$$

Notice that the summation over s_1 has produced a term involving s_2 that wasn't present before – thus, we can say that nibbling off the spin s_1 has caused it to jump to the next site to the right, to become s_2. It has also produced the multiplicative factor $(2\sinh K)$. We show this schematically in Figure 4.5, where the open circle around a site indicates the presence of that spin in the summation.

To complete our calculation, we can now write the average value of s_1 as

$$\langle s_1 \rangle = \frac{1}{Z_2}(2\sinh K)\sum_{\{s_2\}} s_2.$$

The remaining summation (nibble) is easily evaluated as follows:

$$\sum_{\{s_2\}} s_2 = 1 - 1 = 0.$$

This result is valid for any spin. Thus, we find

$$\langle s_1 \rangle = 0. \tag{4.7}$$

This is the expected result.

Next, let's calculate the spin–spin correlation $\langle s_1 s_2 \rangle$ in this two-spin system. This correlation is

$$\langle s_1 s_2 \rangle = \frac{1}{Z_2}\sum_{\{s_1\}}\sum_{\{s_2\}} s_1 s_2 e^{Ks_1 s_2}.$$

Carrying out the summation over s_1 yields

$$\langle s_1 s_2 \rangle = \frac{1}{Z_2}(2\sinh K)\sum_{\{s_2\}} s_2^2.$$

Noting that $s_i^2 = 1$ for any Ising spin, we find

Figure 4.6 When the spin on site 1 jumps to site 2, which already contains a spin, the spin on site 2 is squared (double circle). In this case, the summation of site 2 yields a factor of 2.

$$\langle s_1 s_2 \rangle = \frac{1}{Z_2} (2 \sinh K) \sum_{\{s_2\}} 1 = \frac{2(2 \sinh K)}{2(2 \cosh K)} = \tanh K. \qquad (4.8)$$

We illustrate this graphically in Figure 4.6. When spin s_1 is nibbled off, a factor of $(2 \sinh K)$ is produced, and the spin hops to the next site to the right, where there is already a spin s_2. When this happens, we now have s_2 squared – indicated by the double open circle – which means the spins have effectively "annihilated" one another, since a spin squared is simply equal to 1. As a result, the next nibble produces a factor of 2. The net result is $\langle s_1 s_2 \rangle = \tanh K$. Notice that the spin–spin correlation is zero when the coupling is zero, $K \to 0$, and approaches ± 1 as the coupling goes to plus or minus infinity, $K \to \pm \infty$.

This calculation can be extended to arbitrary separations between the two spins. Each nibble of a site without a spin multiplying the Boltzmann factor generates a factor $(2 \cosh K)$. Each nibble of a site with a spin multiplying the Boltzmann factor yields a factor $(2 \sinh K)$, and the spin hops one site to the right. When the hopping spin encounters the second spin in the correlation, they annihilate one another and the nibbles go back to producing factors of $(2 \cosh K)$ in the interior of the chain, and a factor of (2) at the end of the chain. The partition function in the denominator has factors of $(2 \cosh K)$ for each bond. The net result, then, is that the correlation between spins n sites apart is

$$\langle s_i s_{i+n} \rangle = (\tanh K)^n. \qquad (4.9)$$

This result is independent of the location of the spin s_i. A specific example is given for $n = 2$ in Figure 4.7.

Correlation Length

We now turn our attention to the dependence of spin–spin correlations on the separation between spins. Figure 4.8 shows a series of plots of $\langle s_i s_{i+n} \rangle$ as a function of n for various values of the coupling K. Larger values of K produce

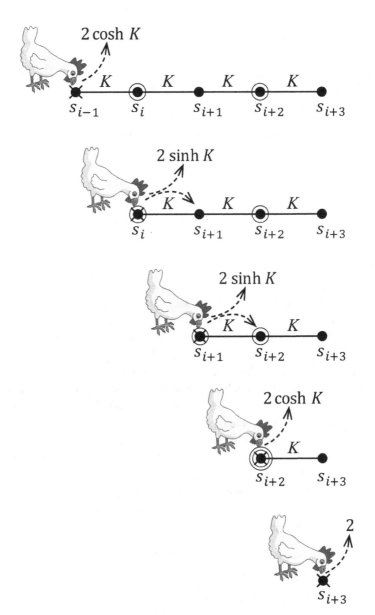

Figure 4.7 Calculating the spin–spin correlation for $n = 2$.

larger values of the correlation for any given n, but there is always a systematic decay of the correlation with increasing separation. In fact, since $(\tanh K)$ is less than 1 for finite values of K, it follows that $\langle s_i s_{i+n} \rangle = (\tanh K)^n$ decreases to zero in the limit $n \to \infty$.

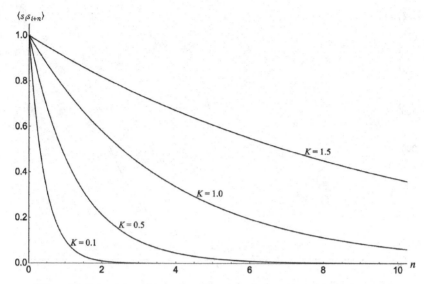

Figure 4.8 Dependence of $\langle s_i s_{i+n} \rangle$ as a function of n for various values of K. In this plot, n is treated as a continuous variable, though on the actual lattice n takes on only integer values.

The precise mathematical form of the decay in spin–spin correlation with distance can be seen from the following key observation: If the separation between spins is increased by an *additive* amount, say from n to $n + 1$, the correlation decreases by a *multiplicative* amount, by $(\tanh K)$ in this case. Specifically,

$$\langle s_i s_{i+n+1} \rangle = (\tanh K)^{n+1} = (\tanh K)(\tanh K)^n = (\tanh K)\langle s_i s_{i+n} \rangle. \quad (4.10)$$

This form of dependence is a hallmark of *exponential* behavior.

Thus, we can write the spin–spin-correlation function in exponential form as follows:

$$\langle s_i s_{i+n} \rangle = (\tanh K)^n = e^{-na/\xi}. \quad (4.11)$$

In this expression, a is the distance between lattice sites, and ξ is the *correlation length*. To verify our observation here, note that

$$\langle s_i s_{i+n+1} \rangle = e^{-(n+1)a/\xi} = e^{-a/\xi} e^{-na/\xi} = e^{-a/\xi} \langle s_i s_{i+n} \rangle.$$

This is in complete agreement with Equation 4.10 if we make this identification:

$$e^{-a/\xi} = \tanh K.$$

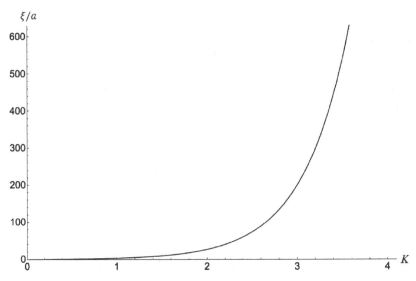

Figure 4.9 Correlation length, ξ, in units of the lattice spacing, a, as a function of K.

Rearranging, and solving for the correlation length, we find

$$\xi = \frac{a}{|\ln(\tanh K)|}.$$ (4.12)

The absolute value sign is used in the denominator to give a positive length, because $(\tanh K)$ is less than 1, and hence $\ln(\tanh K)$ is negative.

We plot ξ as a function of K in Figure 4.9. Notice that the correlation length is zero at zero coupling, as one would expect, and diverges to infinity with infinite coupling. We can think of the correlation length as setting the length scale for the thermodynamic system – that is, the lattice spacing a sets the length scale for the lattice itself, but the correlation length ξ sets the scale over which spins act in a cooperative fashion. In two dimensions, we shall find that the correlation length diverges to infinity as one approaches the critical point, at a finite value of K. Thus, the length scale of spin fluctuations becomes of paramount importance in such systems.

Magnetic Properties

To this point, we've studied the infinite, one-dimensional Ising model in the case of zero magnetic field. Even so, we can still obtain significant results regarding the magnetic properties of the system.

For example, when we add the magnetic field in the next section, the reduced Hamiltonian will contain the following additional term:

$$h \sum_i s_i.$$

Setting $h = 0$ recovers the results of this section. But, before setting h equal to zero, consider taking a derivative of the reduced free energy per site with respect to the magnetic field.

We know from Chapter 2 that the result of this derivative is the average value of the conjugate variable to h; that is, it is equal to the average value of the sum over spins. Thus,

$$\frac{\partial f}{\partial h} = \frac{1}{N} \langle \sum_i s_i \rangle = \frac{1}{N} \sum_i \langle s_i \rangle.$$

Noting that the reduced magnetization, m, is equal to $\partial f / \partial h$, and in addition that the average value of each of the N spins is the same, we have

$$m = \frac{\partial f}{\partial h} = \langle s_i \rangle.$$

As we've seen earlier in this section, specifically in Equation 4.7, the average value of the spins is zero; hence, $m = 0$ in zero field. This result makes it clear that the system is paramagnetic, because only in the ferromagnetic phase is the magnetization nonzero in zero magnetic field.

The next step is to take the second derivative of the free energy with respect to the magnetic field, h. The result, as we've seen in Chapter 2, is the fluctuation of the conjugate variable; that is, the fluctuation in the sum of the spins. In terms of thermodynamics, we refer to this second derivative as the magnetic susceptibility, χ. Therefore,

$$\chi = \frac{\partial^2 f}{\partial h^2} = \frac{1}{N} \left[\langle \left(\sum_i s_i \right) \left(\sum_j s_j \right) \rangle - \langle \left(\sum_i s_i \right) \rangle^2 \right].$$

We've already seen that the second term in this expression is zero, and hence

$$\chi = \frac{1}{N} \langle \left(\sum_i s_i \right) \left(\sum_j s_j \right) \rangle. \tag{4.13}$$

Note that i and j are dummy indices, used for summation purposes. In each of the summations, the sum is over all the spins independently.

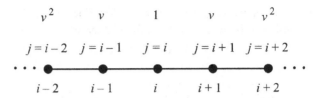

Figure 4.10 Calculating the magnetic susceptibility in Equation 4.14. When the summation index j is equal to i, the calculation generates a contribution of 1. When the index is equal to $i + 1$ or $i - 1$ the contribution is v, where $v = \tanh K$. When the index is equal to $i + 2$ or $i - 2$ the contribution is v^2, and so on.

We can simplify this calculation by noting that all N spins are the same, and hence, rather than summing over all of the spins s_i and then dividing by N, we simply use one of the s_i. Thus,

$$\chi = \langle s_i \sum_j s_j \rangle. \tag{4.14}$$

Figure 4.10 helps to visualize this calculation.

First, let $j = i$. In this case, the average value is the average of s_i^2, which is simply equal to 1. Next, let $j = i - 1$ and $j = i + 1$. The average value in these cases is

$$\langle s_i s_{i-1} \rangle = \langle s_i s_{i+1} \rangle = \tanh K.$$

Letting $v = \tanh K$ for simplicity, we have to this point in the calculation

$$\chi = 1 + 2v + \dots.$$

Next, we let $j = i - 2$ and $j = i + 2$, each of which gives an average value of v^2. Continuing with this procedure, we find

$$\chi = 1 + 2v + 2v^2 + 2v^3 + \dots = 1 + 2v(1 + v + v^2 + \dots).$$

The infinite sum in parentheses is known as a geometric series, and it is easy to evaluate. To see how, let

$$S = 1 + v + v^2 + \dots.$$

Next, note that if we multiply S by v, and then add 1, we are back to the original series. That is,

$$1 + vS = S.$$

This statement is *only* true in the case of an *infinite* series; and hence, we are taking advantage of the infinite nature of the system in writing this equality. Rearranging, we find

$$S = \frac{1}{1-v}.$$

Using this result, we now have

$$\chi = 1 + 2v(1 + v + v^2 + ...) = 1 + \frac{2v}{1-v} = \frac{1+v}{1-v}.$$

Returning to our original notation, with $v = \tanh K$, we have

$$\chi = \frac{1 + \tanh K}{1 - \tanh K}. \tag{4.15}$$

This is certainly a nice, compact expression for the susceptibility. If we express $(\tanh K)$ in terms of exponentials, however, we find an even simpler expression:

$$\chi = \frac{1 + \left(\frac{e^K - e^{-K}}{e^K + e^{-K}}\right)}{1 - \left(\frac{e^K - e^{-K}}{e^K + e^{-K}}\right)} = e^{2K}. \tag{4.16}$$

This result is plotted in Figure 4.11. Notice that it diverges in the limit of infinite coupling, $K \to \infty$ (i.e., $T \to 0$). This means that the spin fluctuations are infinitely large in that limit and indicates a phase transition at $T = 0$. In higher-dimensional systems, χ will diverge at a *finite* temperature; namely, the temperature of the critical point.

It's interesting to note that we've obtained the zero-field magnetic susceptibility – which is the second derivative of the reduced free energy with respect to the magnetic field – even though our calculation is strictly for the case of zero magnetic field. In the next section, we'll take the more conventional approach; we'll first calculate the reduced free energy for finite field, and then take its second derivative with respect to the field.

4.2 The Transfer-Matrix Approach

We turn now to the transfer-matrix method. This is the method that Ising used in his original solution of the one-dimensional finite-field Ising model, and it was also used by Onsager in his solution of the zero-field, two-dimensional Ising model.

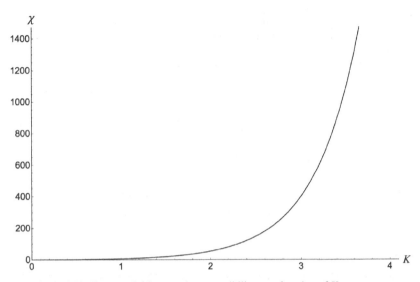

Figure 4.11 The zero-field magnetic susceptibility as a function of K.

The transfer matrix gets its name from the idea that its purpose is to "transfer" the system from one element to another. For example, in the Ising model the transfer matrix takes the system from one site to the next. In optics, the transfer matrix connects the behavior of one optical element to another, and in quantum mechanics the transfer matrix propagates the wave function from one discontinuity in the potential to another. In all of these cases, the basic properties of matrices provide powerful methods for analysis.

Zero Field, Free Boundary Conditions

First, let's consider the Ising model with zero magnetic field, and free boundary conditions. Now, it is often stated – incorrectly – that the transfer-matrix method doesn't apply to free boundary conditions. While it is true that the transfer-matrix method is more effective in systems with periodic boundary conditions, it can be applied to free boundary conditions as well, and we gain deeper insight into the method by doing so.

We start with the simple three-spin system shown in Figure 4.12. The reduced Hamiltonian for this system is

$$-\beta H = K(s_1 s_2 + s_2 s_3).$$

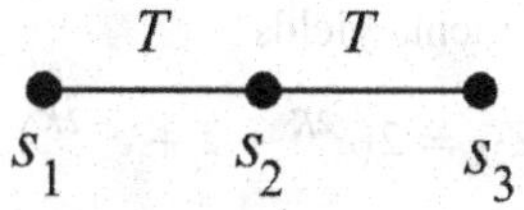

Figure 4.12 A three-spin Ising system with free boundary conditions. This system has two bonds, and one transfer matrix T for each bond.

The partition function is given in Equation 4.2 with $N = 3$:

$$Z_3 = 2^3(\cosh K)^2 = 2(e^{2K} + 2 + e^{-2K}).$$

Recall that this result was obtained with the Nibble approach.

In the transfer-matrix approach, we begin by defining the elements of the transfer matrix in terms of Boltzmann factors as follows:

$$T(s_i, s_j) = e^{Ks_i s_j}. \tag{4.17}$$

The T on the left in Figure 4.12 connects spins s_1 and s_2; the T on the right connects the spins s_2 and s_3. Writing out the transfer matrix in a 2×2 form yields

$$T = \begin{pmatrix} e^K & e^{-K} \\ e^{-K} & e^K \end{pmatrix}. \tag{4.18}$$

This is the zero-field transfer matrix for the one-dimensional Ising model.

We can write the partition function for this system, with free boundary conditions (denoted with the subscript f), as follows:

$$Z_{3f} = \sum_{\{s_1\}} \sum_{\{s_2\}} \sum_{\{s_3\}} e^{Ks_1 s_2} e^{Ks_2 s_3} = \sum_{\{s_1\}} \sum_{\{s_2\}} \sum_{\{s_3\}} T(s_1, s_2) T(s_2, s_3).$$

If we first carry out the summation over s_2, the result is the same as multiplying the two T matrices together. Thus,

$$Z_{3f} = \sum_{\{s_1\}} \sum_{\{s_3\}} T^2(s_1, s_3).$$

This double summation is nothing more than the sum of all four elements of the matrix T^2. We calculate T^2 as follows:

$$T^2 = TT = \begin{pmatrix} e^K & e^{-K} \\ e^{-K} & e^K \end{pmatrix} \begin{pmatrix} e^K & e^{-K} \\ e^{-K} & e^K \end{pmatrix} = \begin{pmatrix} e^{2K} + e^{-2K} & 2 \\ 2 & e^{2K} + e^{-2K} \end{pmatrix}.$$
$$\tag{4.19}$$

Summing all the matrix elements yields

$$Z_{3f} = 2(e^{2K} + 2 + e^{-2K}). \tag{4.20}$$

This result agrees with Equation 4.2, as expected. We can extend this method to larger chains, if desired, but the calculation of T^N makes the approach unwieldy. The transfer-matrix method is simplified by using periodic boundary conditions, which we study next.

Zero Field, Periodic Boundary Conditions

Consider the three-spin Ising model shown in Figure 4.13. Notice that this system has periodic boundary conditions, which means that it has a bond connecting the last spin, s_3, with the first spin, s_1. As a result, the system now requires three transfer matrices, as opposed to just two in the free-boundary-condition case.

For purposes of comparison, we first solve this system with a straightforward summation of the eight configurations of the spins. The reduced Hamiltonian is

$$-\beta H = K(s_1 s_2 + s_2 s_3 + s_3 s_1).$$

Using the subscript p to denote periodic boundary conditions, and summing over the eight spin configurations of the system, we find that the partition function is

$$Z_{3p} = 2(e^{3K} + 3e^{-K}). \tag{4.21}$$

The term e^{3K} corresponds to all spins + (or all spins −), and the terms involving e^{-K} correspond to two spins +, and one spin − (or two spins − and one spin +).

Now, for a different way to solve the problem, we apply the transfer-matrix approach. In this case, the partition function is

$$Z_{3p} = \sum_{\{s_1\}} \sum_{\{s_2\}} \sum_{\{s_3\}} T(s_1, s_2) T(s_2, s_3) T(s_3, s_1).$$

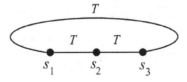

Figure 4.13 A three-spin Ising system with periodic boundary conditions. This system has three bonds, each with a transfer matrix T.

Start by summing over the configurations of s_2 and then summing over the configurations of s_3. The result is

$$Z_{3p} = \sum_{\{s_1\}} \sum_{\{s_3\}} T^2(s_1, s_3) T(s_3, s_1) = \sum_{\{s_1\}} T^3(s_1, s_1).$$

The last summation, where the two indices of the matrix are the same, is over the diagonal elements of T^3. This summation is referred to as the *trace* of the matrix, and it is abbreviated as Tr. Thus, to find the partition function with periodic boundary conditions, we raise T to the third power – one power for each bond – and then calculate its trace.

We've already calculated T^2 in Equation 4.19, so let's multiply that result by T to get the third power:

$$T^3 = T^2 T = \begin{pmatrix} e^{2K} + e^{-2K} & 2 \\ 2 & e^{2K} + e^{-2K} \end{pmatrix} \begin{pmatrix} e^K & e^{-K} \\ e^{-K} & e^K \end{pmatrix}$$
$$= \begin{pmatrix} e^{3K} + 3e^{-K} & 3e^K + e^{-3K} \\ 3e^K + e^{-3K} & e^{3K} + 3e^{-K} \end{pmatrix}. \tag{4.22}$$

Calculating the trace yields

$$Z_{3p} = \mathrm{Tr}\, T^3 = 2(e^{3K} + 3e^{-K}). \tag{4.23}$$

This agrees with the result from direct summation, but it also reveals the potential for a powerful, new approach to the problem.

Diagonalizing the Transfer Matrix

To pursue this alternative approach, we start by determining the eigenvalues of the transfer matrix. This can be achieved with the following calculation:

$$\begin{vmatrix} e^K - \lambda & e^{-K} \\ e^{-K} & e^K - \lambda \end{vmatrix} = (e^K - \lambda)^2 - e^{-2K} = 0.$$

Rearranging and taking the square root, we find

$$(e^K - \lambda) = \pm e^{-K}.$$

Choosing each of the two signs, one at a time, yields

$$\lambda_+ = e^K + e^{-K} = 2\cosh K \tag{4.24a}$$

$$\lambda_- = e^K - e^{-K} = 2\sinh K. \tag{4.24b}$$

Clearly, $\lambda_+ > \lambda_-$ for all values of K.

A similarity transformation generates the diagonal version of the transfer matrix. Specifically,

$$\widetilde{T} = STS^{-1} = S\begin{pmatrix} e^K & e^{-K} \\ e^{-K} & e^K \end{pmatrix}S^{-1} = \begin{pmatrix} \lambda_+ & 0 \\ 0 & \lambda_- \end{pmatrix}. \tag{4.25}$$

The precise form of the similarity matrix S can be determined, but the key point for our purposes is that the trace of a matrix is preserved under a similarity transformation. That is,

$$\mathrm{Tr}\,T = \mathrm{Tr}\,\widetilde{T} = \lambda_+ + \lambda_-.$$

Now, recall that the partition function for the system in Figure 4.13 is the trace of T^3. This trace is unaffected by a similarity transformation, and hence

$$\mathrm{Tr}\,T^3 = \mathrm{Tr}\,ST^3S^{-1}. \tag{4.26}$$

The quantity ST^3S^{-1} in the preceding equation can be rewritten by inserting identity matrices, in the form $I = S^{-1}S$, between the T matrices to yield

$$ST^3S^{-1} = STTTS^{-1} = STS^{-1}STS^{-1}STS^{-1} = \widetilde{T}^3.$$

Substituting this result into Equation 4.26 gives

$$\mathrm{Tr}\,T^3 = \mathrm{Tr}\,ST^3S^{-1} = \mathrm{Tr}\,\widetilde{T}^3. \tag{4.27}$$

The real payoff in performing these transformations is that $\widetilde{T}^3$ has a very simple form. Specifically,

$$\widetilde{T}^3 = \begin{pmatrix} \lambda_+ & 0 \\ 0 & \lambda_- \end{pmatrix}\begin{pmatrix} \lambda_+ & 0 \\ 0 & \lambda_- \end{pmatrix}\begin{pmatrix} \lambda_+ & 0 \\ 0 & \lambda_- \end{pmatrix}$$
$$= \begin{pmatrix} \lambda_+ & 0 \\ 0 & \lambda_- \end{pmatrix}\begin{pmatrix} \lambda_+^2 & 0 \\ 0 & \lambda_-^2 \end{pmatrix} = \begin{pmatrix} \lambda_+^3 & 0 \\ 0 & \lambda_-^3 \end{pmatrix}.$$

Thus,

$$Z_{3p} = \mathrm{Tr}\,T^3 = \mathrm{Tr}\,\widetilde{T}^3 = \lambda_+^3 + \lambda_-^3. \tag{4.28}$$

Substituting for λ_+ and λ_- from Equations 4.24a and 4.24b, we find

$$Z_{3p} = (e^K + e^{-K})^3 + (e^K - e^{-K})^3 = 2(e^{3K} + 3e^{-K}). \tag{4.29}$$

Thus, we recover the result from direct summation given in Equation 4.21.

Taking It to the Limit

Clearly, the preceding results can be generalized to arbitrary N by simply replacing the power 3 with the power N in the previous expressions. In fact, this is the beauty of the procedure – it makes solving for general N a trivial matter. Thus, the partition function for a ring (periodic boundary conditions) of N Ising spins is

$$Z_{Np} = \operatorname{Tr} T^N = \operatorname{Tr} \widetilde{T}^N = \lambda_+{}^N + \lambda_-{}^N. \tag{4.30}$$

No such simplification occurs with free boundary conditions, which depends on the sum of all the matrix elements rather than on the trace. This is why the transfer matrix method is most powerful when used with periodic boundary conditions.

The reduced free energy per site with periodic boundary conditions, f_{Np}, is given by the log of the partition function divided by N. That is,

$$\begin{aligned} f_{Np} &= \frac{1}{N} \ln Z_{Np} = \frac{1}{N} \ln(\lambda_+{}^N + \lambda_-{}^N) = \frac{1}{N} \ln \lambda_+{}^N + \frac{1}{N} \ln\left[1 + \left(\frac{\lambda_-}{\lambda_+}\right)^N\right] \\ &= \ln \lambda_+ + \frac{1}{N} \ln\left[1 + \left(\frac{\lambda_-}{\lambda_+}\right)^N\right]. \end{aligned}$$

$$\tag{4.31}$$

Note that the quantity λ_-/λ_+ is always less than one, and hence when it is raised to the Nth power, with $N \to \infty$ in the thermodynamic limit, it approaches zero. This means that the second term in the above expression vanishes in this limit. Therefore, the reduced free energy per site for an infinite lattice is

$$\begin{aligned} f &= \lim_{N \to \infty} f_{Np} = \ln \lambda_+ + \lim_{N \to \infty} \frac{1}{N} \ln\left[1 + \left(\frac{\lambda_-}{\lambda_+}\right)^N\right] \\ &= \ln \lambda_+ = \ln 2 + \ln(\cosh K). \end{aligned} \tag{4.32}$$

This result agrees with our earlier results for the infinite lattice with free boundary conditions, given in Equation 4.4. In the thermodynamic limit, the system has the same free energy per site, regardless of the boundary conditions.

The significant takeaway from this derivation is that the reduced free energy per site for an infinite system has a very simple form. First, calculate the eigenvalues of the transfer matrix. Second, determine the eigenvalue that is the largest. Third, calculate the free energy by taking the logarithm of the largest eigenvalue. That's all there is to it.

The Finite System with Periodic Boundary Conditions

Let's return to a finite system of N sites with periodic boundary conditions. If we replace the eigenvalues in Equation 4.30 with their expressions in terms of hyperbolic functions, using Equations 4.24a and 4.24b, we have

$$Z_{Np} = (2 \cosh K)^N + (2 \sinh K)^N. \tag{4.33}$$

Note that this expression is valid for all N, and not just for the thermodynamic limit.

A slightly different way of writing the expression in Equation 4.33 is as follows:

$$Z_{Np} = 2^N (\cosh K)^N (1 + v^N). \tag{4.34}$$

In this expression we've used the abbreviation $v = \tanh K$. We shall see this result again in Chapter 5, when we consider series expansions of the Ising model.

The corresponding reduced free energy per site for a finite periodic system is

$$f_{Np} = \frac{1}{N} \ln Z_{Np} = \ln 2 + \ln(\cosh K) + \frac{1}{N} \ln(1 + v^N). \tag{4.35}$$

Using this expression, we can calculate thermodynamic quantities of interest. For example, the reduced specific heat per site is given by K^2 times the second derivative of the free energy with respect to K. That is,

$$c_{Np} = K^2 \frac{\partial^2 f_{Np}}{\partial K^2}. \tag{4.36}$$

The expression for the specific heat is a bit messy to write out explicitly for arbitrary N, but we can learn a lot about the system from plots for various values of N. Specifically, consider Figure 4.14, where we show plots of the specific heat for $N = 5$, 15, 45, and 135. Also shown for comparison, with a dashed curve, is the specific heat for the infinite system. As can be seen, the evolution of the specific heat from finite N to the case of an infinite system is rather interesting and unexpected. This is in contrast to the case of free boundary conditions (Figure 4.4), where the finite-lattice specific heats merge smoothly with the specific heat of the infinite lattice as the number of sites is increased.

As an example of this odd behavior, notice that the peak of the periodic-lattice specific heat doesn't correspond to the peak of the infinite system – that is, the peak of the finite system doesn't merge into the peak of the infinite system as $N \to \infty$. In fact, the peak of the finite system rises above the

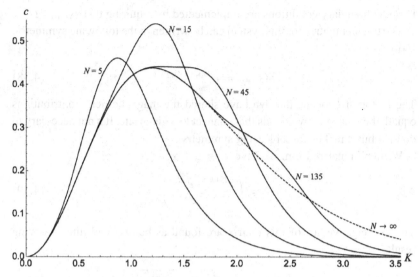

Figure 4.14 Reduced specific heats for finite Ising systems of N sites with periodic boundary conditions. The dashed curve is the result for the infinite lattice.

infinite-lattice peak initially, and then moves to larger values of K where it drops down, flattens out, and merges with the large-K tail of the infinite-system specific heat. The peak of the infinite lattice is actually produced by a "bend" that forms in the small-K portion of the finite-system specific heat; this bend becomes sharper with increasing N and eventually forms a peak in the same location as in the infinite lattice. The result is a specific heat with a decidedly unorthodox shape for large, but finite, values of N.

We often think of periodic boundary conditions as being a better approximation to an infinite system than free boundary conditions, since periodic lattices have no "edges." It's wise to be aware of possibilities like these, however, and to not just assume that periodic boundary conditions will give better results.

Finite Magnetic Field

The power of the transfer-matrix method is illustrated by the fact that in one dimension it applies to both zero and finite magnetic fields. Consider the following reduced Hamiltonian, which includes a term for the magnetic field, h:

$$-\beta H = K \sum_{i=1}^{N} s_i s_{i+1} + h \sum_{i=1}^{N} s_i. \qquad (4.37)$$

Periodic boundary conditions are implemented by requiring that $s_{N+1} = s_1$.

The transfer matrix for this system can be written in the following symmetric form:

$$T(s_i, s_j) = e^{Ks_is_j + \frac{1}{2}h(s_i + s_j)}. \tag{4.38}$$

The factor of $\frac{1}{2}$ means that we have shared the magnetic-field contributions equally between the two bonds that connect to a given site. It's not necessary to do this, but it makes for a pleasing symmetry.

Written in matrix form, we have

$$T = \begin{pmatrix} e^{K+h} & e^{-K} \\ e^{-K} & e^{K-h} \end{pmatrix}. \tag{4.39}$$

The two eigenvalues of this matrix are found as before, with the following results:

$$\lambda_+ = e^K \cosh h + \sqrt{e^{-2K} + e^{2K}(\sinh h)^2}$$

$$\lambda_- = e^K \cosh h - \sqrt{e^{-2K} + e^{2K}(\sinh h)^2}. \tag{4.40}$$

The largest eigenvalue is λ_+, just as it was in zero field. Hence, the reduced free energy per site is

$$f = \ln \lambda_+ = \ln \left[e^K \cosh h + \sqrt{e^{-2K} + e^{2K}(\sinh h)^2} \right]. \tag{4.41}$$

This result is valid for an infinite system, and for all values of K and h.

Finite-Field Magnetization

Thermodynamic quantities are obtained by taking various derivatives of the reduced free energy, f. For example, the magnetization m is given by the first derivative of the free energy with respect to the magnetic field, h. Thus,

$$m = \frac{\partial f}{\partial h} = \frac{e^K \sinh h}{\sqrt{e^{-2K} + e^{2K}(\sinh h)^2}}. \tag{4.42}$$

Notice that m is equal to zero for all finite values of K when $h = 0$, as expected. This means that the system is paramagnetic for all finite K.

For finite magnetic fields, the results are given in Figure 4.15. The rightmost curve corresponds to $K = 0$, which is the limit of independent spins. From our

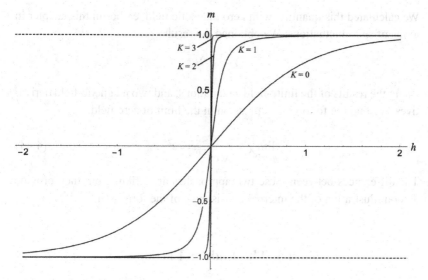

Figure 4.15 Magnetization per site, m, for the infinite one-dimensional Ising system.

discussion in Chapter 2, and Equation 2.8 in particular, we know that $m = \tanh h$ in this limit. This result also agrees with Equation 4.42 when K is set equal to zero.

When K is finite, the spins are coupled. As the coupling increases, which corresponds to the temperature going toward zero, the magnetization curves have a sharp bend near $h = 0$. This means that the coupling keeps the spins aligned until the magnetic field is almost zero. When the magnetic field passes through zero, the sign of the magnetization suddenly flips. In the limit of infinite coupling – or zero temperature – the magnetization remains at +1 as the magnetic field is reduced to zero, at which point it jumps discontinuously to −1 as the magnetic field becomes negative. This shows that the system is ferromagnetic only in the limit of $T \to 0$.

Magnetic Susceptibility

One more derivative of the free energy f with respect to h gives the magnetic susceptibility:

$$\chi = \frac{\partial^2 f}{\partial h^2} = \frac{e^K \cosh h}{[1 + e^{4K}(\sinh h)^2]\sqrt{e^{-2K} + e^{2K}(\sinh h)^2}}. \qquad (4.43)$$

We calculated this quantity, with zero magnetic field, earlier in this chapter in terms of spin fluctuations. We obtained the result

$$\chi = e^{2K}.$$

Using the results of the finite-field calculations, and two magnetic-field derivatives, we find the following expression in the limit of zero field:

$$\lim_{h \to 0} \chi = \frac{e^K}{\sqrt{e^{-2K}}} = e^{2K}. \tag{4.44}$$

The differences between these two approaches are striking, but they provide a good illustration of the internal consistency of the formalism.

4.3 The Ladder Lattice

A natural extension of the one-dimensional $1 \times \infty$ lattice is a $2 \times \infty$ lattice, like the one shown in Figure 4.16 (a). This lattice has two spins in the vertical

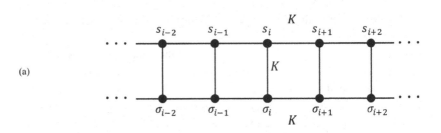

(a)

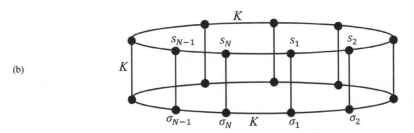

(b)

Figure 4.16 The ladder lattice. (a) The nearest-neighbor couplings K and spin-labeling system for the infinite $2 \times \infty$ ladder lattice. (b) A finite $2 \times N$ ladder lattice with periodic boundary conditions.

direction, and extends to infinity in the horizontal direction. Because of its
shape, it is often referred to as a ladder lattice.

For the sake of simplicity, we'll consider only nearest-neighbor couplings K between the spins, as indicated in Figure 4.16 (a). In addition, to
help with notation, we'll label the top row of spins with s_i and the bottom
row with σ_i. The elements of the transfer matrix T, which takes us from the
spins s_i and σ_i to the spins s_{i+1} and σ_{i+1}, can be written as follows:

$$T_{(s_i,\sigma_i;s_{i+1},\sigma_{i+1})} = e^{K\left[s_i s_{i+1}+\sigma_i\sigma_{i+1}+\frac{1}{2}(s_i\sigma_i+s_{i+1}\sigma_{i+1})\right]}. \tag{4.45}$$

A factor of one half multiplies the terms corresponding to vertical bonds,
because these bonds are shared equally between adjacent units of the
lattice. We also employ periodic boundary conditions, as indicated in
Figure 4.16 (b).

Writing out the transfer matrix, we have the following:

$$T = \begin{pmatrix} T_{(1,1;1,1)} & T_{(1,1;1,-1)} & T_{(1,1;-1,1)} & T_{(1,1;-1,-1)} \\ T_{(1,-1;1,1)} & T_{(1,-1;1,-1)} & T_{(1,-1;-1,1)} & T_{(1,-1;-1,-1)} \\ T_{(-1,1;1,1)} & T_{(-1,1;1,-1)} & T_{(-1,1;-1,1)} & T_{(-1,1;-1,-1)} \\ T_{(-1,-1;1,1)} & T_{(-1,-1;1,-1)} & T_{(-1,-1;-1,1)} & T_{(-1,-1;-1,-1)} \end{pmatrix}$$

$$= \begin{pmatrix} e^{3K} & 1 & 1 & e^{-K} \\ 1 & e^K & e^{-3K} & 1 \\ 1 & e^{-3K} & e^K & 1 \\ e^{-K} & 1 & 1 & e^{3K} \end{pmatrix}. \tag{4.46}$$

By using any of the standard mathematics program available today, the four
eigenvalues of this matrix can be obtained. The reduced free energy per site is
given by the log of the largest eigenvalue. We find

$$f = \frac{1}{2}\log\left[(\cosh K)(2\cosh 2K + \sqrt{10 - 8\cosh 2K + 2\cosh 4K})\right]. \tag{4.47}$$

The factor of one half multiplying the log takes account of the fact that two
spins occupy each unit of the lattice. The thermodynamic functions associated
with this free energy are very similar to those already obtained for the $1 \times \infty$
lattice.

2-D or Not 2-D

It's reasonable to think of the ladder lattice as a two-dimensional system. After
all, it extends in both the vertical and horizontal directions. That makes it two
dimensional – or so one might think.

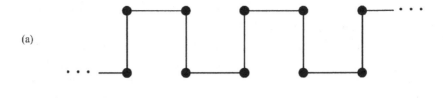

(a)

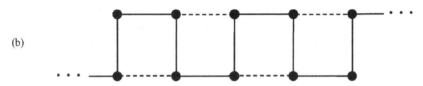

(b)

Figure 4.17 The ladder lattice is a one-dimensional lattice with additional inter-actions. (a) Redraw the 1-D lattice as a zigzag lattice, which has no effect on the thermodynamics. (b) Add interactions between every other third neighbor (dashed lines) on the one-dimensional lattice to obtain the ladder lattice.

To see the issue more clearly, consider redrawing the one-dimensional lattice as shown in Figure 4.17 (a). This change has no effect as far as the thermodynamics of the system are concerned – all of the spins have the same interactions as before. All we've done is drawn the lattice in such a way that it zigs up and zags down instead of extending in a straight line. Drawn this way, the system looks two-dimensional, in a purely geometric sense, but it's clearly one-dimensional in terms of its thermodynamic properties.

If we modify the lattice in Figure 4.17 (a) slightly by adding third neighbor interactions for every other spin, as indicated with dashed lines in Figure 4.17 (b), we obtain the ladder lattice. Thus, the ladder lattice is also a one-dimensional lattice – thermodynamically speaking – because it is simply a standard $1 \times \infty$ lattice with an additional set of interactions.

Another way to look at the situation is to recall that as one approaches a critical point the spin–spin correlation length diverges to infinity. From the perspective of an infinite correlation length, the finite vertical extent of a $2 \times \infty$ lattice is insignificant, and the thermodynamic behavior is the same as if the system were purely one-dimensional.

4.4 The Ladder Lattice with the Nibble Approach

The zero-field ladder lattice with nearest-neighbor coupling K can be solved with a slight generalization of the Nibble approach. The approach also applies to finite fields, but we consider only the zero-field case here for simplicity.

The first step is to nibble off the two spins on the end of the ladder, as indicated with Xs in Figure 4.18. These spins, s_1 and σ_1, are connected only to the spins s_2 and σ_2. We can write the reduced Hamiltonian for these four spins as follows:

$$-\beta H = K_1 s_1 \sigma_1 + K(s_1 s_2 + \sigma_1 \sigma_2 + s_2 \sigma_2). \tag{4.48}$$

This forms the basis for our transformation.

Notice that we've singled out the vertical interaction between s_1 and σ_1 with a different interaction term, K_1. On the first step, we set K_1 equal to K so that all interactions are the same – as in the original system. On the next step, however, we will find that the effective interaction between s_2 and σ_2, which we label K_1', is different from the rest of the interactions on the lattice. As we continue the calculation, each step after the first has a vertical interaction between the end spins that is different from the other interactions. It is for this reason that we derive the transformation for the general case of an arbitrary vertical interaction at the end of the lattice.

The effective reduced Hamiltonian between the spins s_2 and σ_2 is

$$-\beta H_{\text{eff}} = 2K_0' + K_1' s_2 \sigma_2. \tag{4.49}$$

The factor of 2 in front of the K_0' term takes account of the two spins in the effective system; with this factor in place, it follows that K_0' represents the reduced free-energy contribution *per site*.

The two effective quantities, K_0' and K_1', are found with the following calculation:

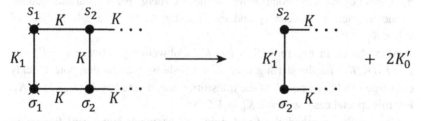

Figure 4.18 The Nibble approach applied to the ladder lattice. Summing s_1 and σ_1 yields an effective coupling between s_2 and σ_2, and produces a contribution of $2K_0'$ to the reduced free energy per site, f.

$$\sum_{\{s_2\}} \sum_{\{\sigma_2\}} e^{2K_0' + K_1' s_2 \sigma_2} = \sum_{\{s_2\}} \sum_{\{\sigma_2\}} \sum_{\{s_1\}} \sum_{\{\sigma_1\}} e^{K_1 s_1 \sigma_1 + K(s_1 s_2 + \sigma_1 \sigma_2 + s_2 \sigma_2)}. \qquad (4.50)$$

In words, this expression says that summing the effective Boltzmann factors over the configurations of s_2 and σ_2 is the same as summing the original Boltzmann factors over the configurations of s_1, σ_1, s_2, and σ_2. This guarantees that the partition function is preserved.

To carry out this calculation, assign values to s_2 and σ_2, and then sum over the four configurations of s_1 and σ_1. For example, suppose $s_2 = +1$ and $\sigma_2 = +1$. In this case, we find that the partial partition function, Z_{++}, is given by

$$Z_{++} = e^{2K_0' + K_1'} = e^K (e^{-2K + K_1} + e^{2K + K_1} + 2e^{-K_1}) = e^K Z_{++}'. \qquad (4.51)$$

Similarly, with $s_2 = +1$ and $\sigma_2 = -1$ we have the partial partition function Z_{+-}:

$$Z_{+-} = e^{2K_0' - K_1'} = e^{-K} (e^{-2K - K_1} + e^{2K - K_1} + 2e^{K_1}) = e^{-K} Z_{+-}'. \qquad (4.52)$$

The up–down symmetry of the original system guarantees that these are the only two independent partial partition functions. Solving these equations for K_0' and K_1' yields

$$K_0'(K, K_1) = \frac{1}{4} \ln Z_{++} Z_{+-} = \frac{1}{4} \ln Z_{++}' Z_{+-}'$$

$$K_1'(K, K_1) = \frac{1}{2} \ln \frac{Z_{++}}{Z_{+-}} = K + \frac{1}{2} \ln \frac{Z_{++}'}{Z_{+-}'}. \qquad (4.53)$$

These equations define the Nibble transformation for this lattice.

Notice that $K_1' = K$ for $K_1 = 0$, since the primed partial partition functions are equal in this limit. We can see this result in Figure 4.19, where we plot $y = K_1'(K, K_1)$ as a function of K_1 for the special case $K = 1.0$. Physically, the meaning of this result is that, if $K_1 = 0$ in Figure 4.18, then there is no "through line of communication" from s_2 to s_1 to σ_1 to σ_2. As a result, summing over s_1 and σ_1 simply generates two contributions to the free energy per site, but adds nothing to the interaction between s_2 and σ_2. It follows that K_1' equals its original value, $K_1' = K$.

Also plotted in Figure 4.19 is $y = K_1$. Cobwebbing between $y = K_1$ and $y = K_1'(K, K_1)$ for the starting value $K = 1.0$ shows that the iterations quickly converge to a "fixed point" of the transformation, defined by $K_1'(K, K_1^*) = K_1^*$. For this special case, we have $K_1^* = 1.596 \ldots$.

Once we've reached the fixed point, to as many significant figures as desired, the remaining steps of the calculation continue to produce the same

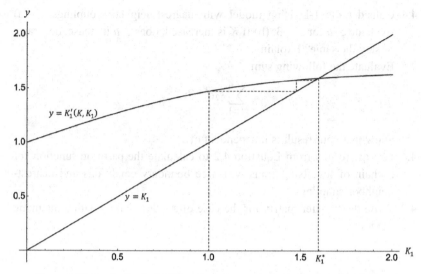

Figure 4.19 Results for the Nibble calculation on the ladder lattice. The intersection of $y = K_1$ and $y = K_1'(K, K_1)$ yields the fixed point K_1^*. In this case, $K = 1.0$ and $K_1^* = 1.596 \ldots$.

value for K_1' – and this will occur an infinite number of times. Thus, the reduced free energy per site f is simply the function K_0' evaluated at the fixed point:

$$f = K_0'(K, K_1^*).$$

With $K = 1.0$ and $K_1^* = 1.596 \ldots$ we find $K_0' = 1.514 \ldots$, which agrees with the exact free energy f given in Equation 4.47 when the coupling is set to the value $K = 1.0$.

4.5 Problems

4.1 In Equation 4.2, explain why 2 is raised to the power N, but $\cosh K$ is only raised to the power $N - 1$.

4.2 (a) Find the reduced entropy per site in the limit $K \to \infty$ for a 1-D lattice with N sites. (b) What is this entropy in the thermodynamic limit? (c) Are your results in parts (a) and (b) different for periodic versus free boundary conditions?

4.3 Referring to Figure 4.9, approximately what value of K yields a correlation length equal to 500 lattice spacings?

4.4 Consider the 1-D Ising model with nearest-neighbor couplings K. (a) Evaluate ξ/a for $K = 3$. (b) If K is increased, does ξ/a increase, decrease, or stay the same? Explain.

4.5 Evaluate the following sum:

$$\sum_{\{s_1\}} s_2 e^{Ks_1s_2}.$$

Show that your result is independent of s_1.

4.6 Use the results from Equation 4.2 to calculate the partition function for a chain of four Ising spins with free boundary conditions and nearest-neighbor couplings K.

4.7 Write the transfer matrix for the case of a ladder lattice with a magnetic field, h.

5

The Onsager Solution and Exact Series Expansions

To this point, we've studied several exact methods of calculation for Ising models. These studies culminated in the exact solution for an infinite one-dimensional Ising model, as well as the corresponding solution on a $2 \times \infty$ lattice. Neither of these systems shows a phase transition, however. In this chapter, we start with Onsager's exact solution for the two-dimensional lattice, which quite famously does have a phase transition. Next, we explore exact series expansions from low and high temperature, and show how these results can be combined, via the concept of duality, to give the exact location of the phase transition in two dimensions. We end with a discussion of universality.

5.1 The Onsager Solution

The single most important development in the study of the Ising model – without question – is the exact solution of the two-dimensional, square lattice, nearest-neighbor, zero-field model obtained by Lars Onsager in 1944. This solution shows that the Ising model does indeed have a phase transition from a high-temperature paramagnetic phase to a low-temperature ferromagnet phase at a finite temperature. The phase transition is marked by a singularity, which is best illustrated by a rather dramatic peak in the specific heat; in fact, the peak of the specific heat diverges to infinity, as we shall see.

Onsager obtained his solution with the transfer-matrix method. In fact, he started by solving the $2 \times \infty$ ladder lattice, just as we did in the previous chapter. He then applied the same techniques to the $3 \times \infty$, $4 \times \infty$, $5 \times \infty$, and $6 \times \infty$ lattices, in each case finding the largest eigenvalue of the transfer matrix, and then taking the logarithm of it to obtain the free energy. At that point he was able to generalize – in what is aptly referred to as a mathematical tour de force – to the case of an $\infty \times \infty$ lattice for the desired 2-D solution.

The calculation performed by Onsager is beyond the scope and length restrictions of this presentation, but we can give the final results. Even doing that is a bit challenging, though, because they depend on what are known as elliptic integrals, which cannot be solved analytically. Thus, numerical integration is required to obtain results that can be plotted.

Another important point about the Onsager solution is that it is restricted to a very specific Ising model – an Ising model with nearest-neighbor interactions only, on a two-dimensional square lattice, and with zero magnetic field. To be specific, the reduced Hamiltonian for Onsager's solution is

$$-\beta H = K \sum_{\langle i,j \rangle} s_i s_j.$$

The summation symbol indicates a sum over all nearest neighbor pairs of spins on the square lattice shown in Figure 5.1. No one has been able to extend the solution to longer-range interactions, three dimensions, or finite magnetic fields. Thus, the Onsager solution, as important as it is, is limited in the range of its applicability.

We present the main results from Onsager's solution in this section. We also present plots to illustrate the behavior of the various thermodynamic quantities.

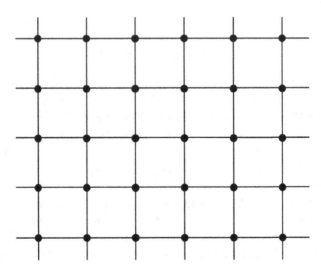

Figure 5.1 The two-dimensional square lattice for the Onsager solution. The model includes only interactions of strength K between nearest-neighbor pairs of spins.

The Free Energy

The reduced free energy per site, f, is the fundamental relation that contains all of the thermodynamic behavior of a system. For the two-dimensional square lattice with nearest-neighbor couplings K, Onsager obtained the following result:

$$f(K) = \frac{1}{2}\ln 2 + \frac{1}{2\pi}\int_0^\pi \ln\left[(\cosh 2K)^2 + \frac{1}{u}\sqrt{1 + u^2 - 2u\cos(2\theta)}\right] d\theta. \quad (5.1)$$

In this expression, the quantity u is given by

$$u = \frac{1}{(\sinh 2K)^2}. \quad (5.2)$$

There is no closed-form, analytic expression for the integral in Equation 5.1. It can be evaluated in various limits, like for large or small values of K, for example, but in general a numerical integration is required.

Figure 5.2 shows f, represented by the solid curve, as a function of the nearest-neighbor coupling K. The first thing we notice about f is how simple it is – it starts at a finite value for small K and gently curves upward to approach a straight line for large K. There would be no way to know – just by looking at this curve – that it represents a system with a phase transition.

Let's look at the small and large K limits more carefully. At high temperature ($K \to 0$) the system is all entropy, as we've seen numerous times before. Thus, the reduced free energy per site in this limit is $f = \ln 2$. At low temperature ($K \to \infty$) the system is all energy. Noting that there are two couplings per site, one vertical and one horizontal, this limit is $f = 2K$.

The vertical dashed line in Figure 5.2 is of particular interest – it represents the location of the phase transition (or critical point) of the Ising model. As we shall see later in this chapter, the location of the critical point, K_c, is given by the following expression:

$$K_c = \frac{1}{2}\ln(1 + \sqrt{2}) = 0.44068\ldots \quad (5.3)$$

You would never know that this is a special value of K by just looking at the free energy. The significance of the critical point becomes increasingly evident, however, as we consider the first and second derivatives of the free energy; that is, as we consider the internal energy and specific heat, respectively.

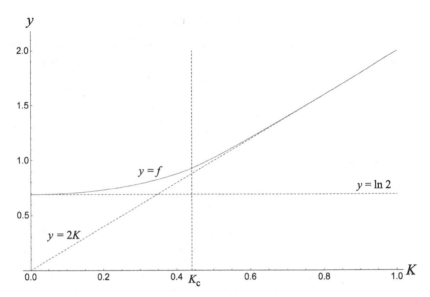

Figure 5.2 The reduced free energy per site, $y = f$, plotted as a function of the nearest-neighbor coupling, K. The horizontal dashed line, $y = \ln 2$, is the high-temperature limit of the free energy; the inclined dashed line, $y = 2K$, is the low-temperature limit; the vertical dashed line is the location of the phase transition (critical point) at $K = K_c = 0.44068$ ....

Internal Energy

The reduced energy per site for this system, which we designate by the symbol e, is defined as follows:

$$e = \langle -\beta H \rangle = \langle K \sum_{\langle i,j \rangle} s_i s_j \rangle = K \langle \sum_{\langle i,j \rangle} s_i s_j \rangle.$$

Note that the average value of the sum over nearest-neighbor spins is equal to the derivative of the free energy with respect to the corresponding conjugate variable, which in this case is K. Thus,

$$e = K \langle \sum_{\langle i,j \rangle} s_i s_j \rangle = K \frac{\partial f}{\partial K}.$$

This quantity can be evaluated by taking the derivative of the integrand in the expression for the free energy. We find

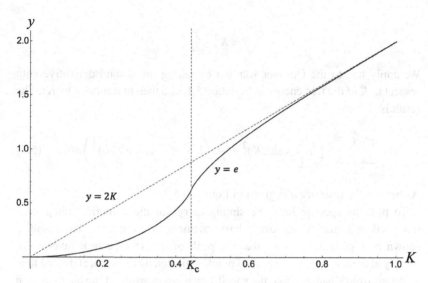

Figure 5.3 The reduced internal energy per site, e, of the two-dimensional Ising model (solid curve) as a function of the nearest-neighbor coupling K, and its large K limit, $2K$ (dashed sloping line). The vertical dashed line indicates the location of the critical point, K_c. At this point the slope of the internal energy is infinite.

$$e = \frac{K}{2\pi} \int_0^\pi \frac{\partial}{\partial K} \left(\ln \left[(\cosh 2K)^2 + \frac{1}{u} \sqrt{1 + u^2 - 2u \cos(2\theta)} \right] \right) d\theta. \quad (5.4)$$

The quantity u is given by Equation 5.2, as before. This integral must be evaluated numerically to yield results suitable for plotting.

Carrying out the numerical integration, we obtain the plot shown in Figure 5.3. Notice that the energy is zero when the coupling K is zero, as expected, and that it approaches $2K$ in the limit of large coupling – that is, in the limit of low temperature.

As before, the vertical dashed line shows the location of the critical point, which is more apparent in this case. In fact, the *slope* of the curve for e diverges at the single point $K = K_c$.

Specific Heat

We turn now to the most dramatic illustration of the critical point in the two-dimensional Ising model – the infinite peak in the reduced specific heat per site, c. To obtain this result, we use the expression first derived in Equation 2.16; namely,

$$c = K^2 \frac{\partial^2 f}{\partial K^2}.$$

We apply this to the Onsager solution by taking the second derivative with respect to K of the free energy in Equation 5.1, and then multiplying by K^2. The result is

$$c = \frac{K^2}{2\pi} \int_0^\pi \frac{\partial^2}{\partial K^2} \left(\ln\left[(\cosh 2K)^2 + \frac{1}{u}\sqrt{1 + u^2 - 2u\cos(2\theta)} \right] \right) d\theta. \qquad (5.5)$$

As before, the quantity u is given in Equation 5.2.

To plot the specific heat, we simply carry out the numerical integration indicated in Equation 5.5 using basic mathematics software. The result is shown in Figure 5.4. Notice that the peak of the specific heat diverges to infinity at the location of the critical point, indicated by the vertical dashed line.

Recall from Chapter 2 that the specific heat is a measure of the *fluctuation* in the energy of the system. The average energy itself is perfectly finite at the critical point, as shown in Figure 5.3, but the fluctuations about that average tend to grow as the transition is approached.

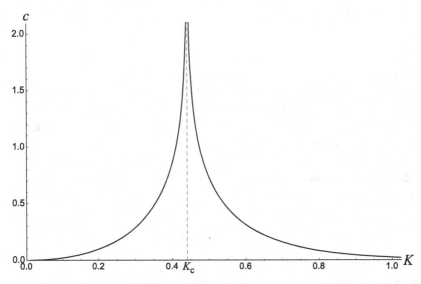

Figure 5.4 The reduced specific heat per site, c, of the two-dimensional Ising model as a function of the nearest-neighbor coupling K. The vertical dashed line is the location of the critical point, K_c. The specific heat goes to infinity at that point.

Here's a way to see how this can come about. Imagine flipping 20 coins and assigning +1 to heads and –1 to tails. The average value obtained by adding up the coins after each one is flipped is zero. The fluctuations about this value are fairly small, on the order of roughly plus or minus 2 or 4, because it's unlikely to flip the coins and have 15 or 20 coins come up heads.

Now, imagine a coupling between the coins, like the coupling between spins in the Ising model. To do this, suppose we take 10 coins and lay them side-by-side, with the same face upward. Now glue them together to form a long strip. This couples them to one another. Do the same to the other 10 coins. Now, when we flip the coins, we can have the following results: 20 heads (sum = +20); 10 heads and 10 tails (sum = 0); or 20 tails (sum = –20). The average value is still zero, as before, but now the fluctuations are as large as the system itself – ranging from +20 to –20.

Something very much like this happens in the Ising system, as illustrated in Figure 5.5. As the critical point is approached from the small K side, shown on the left half of Figure 5.5, clusters of spins pointing in the same direction develop and grow in size as the coupling strengthens. This is indicated by the dark patches on the light (disordered) background. The typical size of the clusters is measured by the correlation length ξ. When large patches of coupled spins flip from one sign to the other, the fluctuation in energy can grow infinitely large as the correlation length goes to infinity. On the other side of the critical point, the dark background indicates spins in the majority direction, and the light patches are clusters of spins pointing in the opposite direction. As before, the fluctuations can be unbounded near K_c.

Another point of interest regarding the peak in the specific heat is the following: The height of the peak is *infinite*, but the area under the peak is *finite*. That is, the peak grows higher and higher, but at the same time it

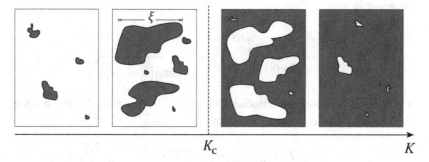

Figure 5.5 Clusters of spins grow in size near the critical point, resulting in infinite fluctuations in the energy of the system, and an infinite peak in the specific heat.

becomes narrower and narrower, resulting in a finite area. The significance of this is seen when one considers that the energy required to raise the temperature of a system from one temperature to another is equal to the corresponding area under the specific heat. Thus, even though the specific heat is infinitely large at the critical point, only a finite amount of energy is required to cross from one side of the critical temperature to the other. If the area under the specific heat had been infinite, as one might at first suppose it to be, then it would be impossible to raise the temperature from below to above the critical value.

Spontaneous Magnetization

The final thermodynamic function we consider is the magnetization with zero magnetic field, m – also referred to as the *spontaneous magnetization* because it arises spontaneously, without a magnetic field to produce it. This result was obtained by Onsager and presented at a conference in 1944, but was first published by C. N. Yang in 1952. The derivation is quite complicated, as one might expect, but remarkable cancellations occur near the end, leading to the following surprisingly simple expression:

$$m = \left[1 - \frac{1}{(\sinh 2K)^4}\right]^{1/8}. \tag{5.6}$$

Note that this expression is valid only for $K > K_c$, which means for temperatures *below* the critical temperature, $T < T_c$. Above the critical temperature, the magnetization is zero; $m = 0$.

A convenient way to plot this result is as a function of temperature, T. To do this, we work with dimensionless quantities; that is, we let $\frac{J}{k_B} = 1$ (which is similar to setting $c = 1$ in relativity, or setting $\hbar = 1$ in quantum mechanics). The result is

$$m = \left[1 - \frac{1}{\left(\sinh\left(\frac{2}{T}\right)\right)^4}\right]^{1/8} = \left[1 - \frac{1}{\left[\sinh\left(\frac{2}{(T/T_c)T_c}\right)\right]^4}\right]^{1/8}. \tag{5.7}$$

In this expression, we have introduced the critical temperature, T_c, which is the inverse of the critical coupling in Equation 5.3. Thus,

$$T_c = \frac{1}{0.44068\ldots} = 2.2691\ldots.$$

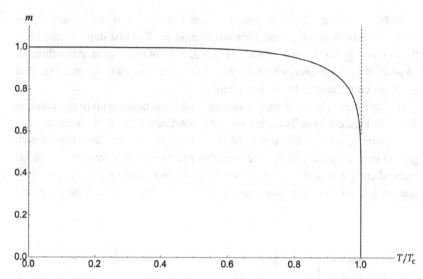

Figure 5.6 Spontaneous (zero-field) magnetization as a function of temperature. The vertical dashed line indicates the location of the critical temperature.

A plot of m as a function of T/T_c is given in Figure 5.6. As before, the vertical dashed line shows the location of the critical temperature. The magnetization is zero for temperatures above the critical temperature.

At the critical temperature, an infinite cluster forms in the system, leading to the infinite fluctuations indicated in the specific heat. This infinite cluster, however, accounts for only a vanishingly small fraction of the spins in the infinite system. As a result, the magnetization starts off at zero just at T_c. Once the temperature is lowered below T_c, the magnetization rapidly saturates to $m \sim 1$ as the infinite cluster spreads and overwhelms the system.

5.2 Low-Temperature Series Expansions

Exact solutions in physics are few and far between – especially for systems with complex features like singularities. The Onsager solution is one of these rare exceptions, but if we want to study other Ising models – say an Ising model in three dimensions, or one with a magnetic field – we must apply other methods. In this section and the next, we look at *exact* series expansions of the reduced free energy. Though these expansions have only a finite number of terms, as opposed to the infinity of terms in a full solution, they have been of great value in advancing our understanding of the Ising model.

To begin, we develop a low-temperature series expansion for the nearest-neighbor, zero-field Ising model in two dimensions. The first step is to calculate the free energy for the ground state ($T = 0$, $K = \infty$) of the system. After that, we consider the lowest-energy excitations above the ground state one at a time, and add their contributions to the free energy.

One of the ground states for our system is all spins pointing up (black dots), as shown in Figure 5.7 (a). There is a second ground state as well, of course, with all spins pointing down. Notice that all spins point in the *same direction* in both ground states, and hence all nearest-neighbor pairs of spins contribute $+K$ to the reduced energy. In addition, recall that there are two nearest neighbors per site – one vertical and one horizontal – and hence if there are N sites on the lattice the

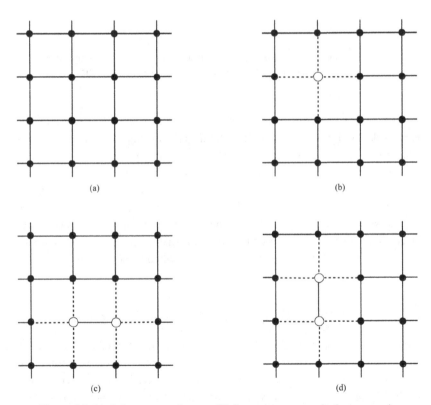

Figure 5.7 (a) Spin-up ground state. (b) Lowest-energy excitation; one spin flipped, four bonds broken. (c) Second-lowest-energy excitation; two nearest-neighbor spins flipped in the horizontal direction, six bonds broken. (d) Another second-lowest-energy excitation; two nearest-neighbor spins flipped in the vertical direction, six bonds broken.

reduced energy is $2NK$. It follows that the partition function for the two ground states is

$$Z_N = 2e^{2NK}.$$

Taking the limit as the number of sites goes to infinity, we find that the ground-state reduced free energy per site is

$$f(K) = \lim_{N \to \infty} \left[\frac{1}{N} \ln Z_N \right] = \lim_{N \to \infty} \left[\frac{1}{N} (\ln 2 + 2NK) \right] = 2K.$$

This limit agrees with the inclined, dashed line in Figure 5.2, which shows the large-K, low-T limit of the exact Onsager free energy.

Next, we consider the lowest-energy excitation away from the ground state. This is illustrated in Figure 5.7 (b), where we see a single spin (shown in white) that is flipped to the opposite direction of the others. Flipping this spin "breaks" the surrounding nearest neighbor bonds (indicated with dashed lines), changing each of them from $+K$ to $-K$. This results in a *change* of the reduced energy of $-2K$ for each of the four broken bonds. The corresponding Boltzmann factor for each broken bond is e^{-2K}, and hence it is $(e^{-2K})^4$ for all four broken bonds taken together.

It follows that the reduced energy of the excited state in Figure 5.7 (b) is

$$2NK + 4(-2K).$$

This excitation occurs N times, corresponding to flipping any one of the N spins. With these results, we can now write the partition function to this order:

$$Z_N = 2e^{2NK} + 2Ne^{2NK+4(-2K)} = 2e^{2NK}[1 + N(e^{-2K})^4].$$

The associated reduced free energy is

$$f(K) = \lim_{N \to \infty} \left[\frac{1}{N} \log Z_N \right] = 2K + (e^{-2K})^4.$$

Notice that we've expanded the log to obtain this result, using the first term in the following expression:

$$\ln(1 + x) = x - \frac{x^2}{2} + \dots.$$

In this case, $x = N(e^{-2K})^4$, which is small for large K and finite N. The limit of $N \to \infty$ is taken after the log is expanded, and hence the N cancels.

Continuing with this procedure, we see that the next-higher-energy excitation from the ground state occurs when we flip two spins that are nearest neighbors of one another, as shown in Figures 5.7 (c) and (d). The first spin can be flipped at any of N sites, and the second spin is a nearest neighbor of the first. Recall, however, that there are just two nearest neighbors per site – one horizontal and one vertical – if we are to avoid double counting. Thus, we have $2N$ of these excitations, each of which breaks six bonds. The partition function is now

$$Z_N = 2e^{2NK}[1 + N(e^{-2K})^4 + 2N(e^{-2K})^6].$$

The reduced free energy to this order is

$$f(K) = \lim_{N \to \infty} \left[\frac{1}{N} \ln Z_N\right] = 2K + (e^{-2K})^4 + 2(e^{-2K})^6.$$

Recall that each term in this expansion is exact.

This basic process can be continued to higher and higher powers of e^{-2K}, though the counting of excited states becomes increasingly complicated. In fact, an oft quoted rule of thumb is that "it takes as much work to calculate the next term in the expansion as it took to calculate all of the preceding terms combined." Put another way, the work required to calculate additional terms increases exponentially. Even so, dozens of terms have been calculated for all sorts of models, and doing these kinds of calculations is an area of specialty in itself.

Specific Heat

Let's do a quick comparison between our series expansion results so far and the Onsager solution. The most dramatic feature in the two-dimensional Ising model, and the best place to test our results, is the reduced specific heat per site,

$$c = K^2 \frac{\partial^2 f}{\partial K^2}.$$

Using our series-expansion result for the reduced free energy, we find

$$c = K^2[64(e^{-2K})^4 + 288(e^{-2K})^6].$$

The next term in the expansion, just for the sake of comparison, is $K^2 1152(e^{-2K})^8$.

We show these results on the large-K side of Figure 5.8. In particular, the lowest dashed curve is the fourth-order term, the dashed curve in the middle is the fourth-order term plus the sixth-order term, and the highest dashed curve is

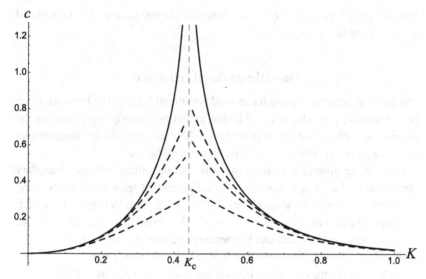

Figure 5.8 The Onsager reduced specific heat per site (solid curve) compared with various series expansion results (dashed curves) from the low-temperature side (large K) and the high-temperature side (small K). The vertical dashed line is the location of the singularity at the critical point.

the sum of the fourth-, sixth-, and eighth-order terms. The solid curve is the exact Onsager solution.

It's clear that our approximations improve with each additional term in the series expansion. It's also clear, however, that the singular behavior at the critical point will be seen only with an infinite number of terms.

Higher Dimensions

To this point, we've considered only the two-dimensional, square lattice, but the same approach can be applied to other lattices as well. For example, suppose we would like to do an expansion on a d-dimensional, hypercubic lattice – that is, a lattice that generalizes from the square lattice ($d = 2$) and simple cubic lattice ($d = 3$) to arbitrary dimension d. The result is

$$f(K) = dK + (e^{-2K})^{2d} + d(e^{-2K})^{2(2d-1)}.$$

You can check that this reproduces our previous result for the case of $d = 2$. If we now apply it to the 3-D simple cubic lattice, $d = 3$, we have

$$f(K) = 3K + (e^{-2K})^6 + 3(e^{-2K})^{10}.$$

The changes in critical behavior as a function of dimension will be considered in the last section of this chapter.

One-Dimensional Behavior

The low-temperature approach can shed additional light on the behavior of an Ising model on a one-dimensional lattice, and in particular on the reason for the absence of a phase transition in that case. The same argument suggests that there *is* a phase transition in the two-dimensional lattice.

The spin-up ground state for a one-dimensional lattice with free boundary conditions is shown in Figure 5.9 (a). Notice that all spins point in the same direction, and that all bonds are "satisfied" – as indicated by the solid connecting lines. The first excited state, with one broken bond (dashed line), is shown in Figure 5.9 (b). The break can happen anywhere on the lattice, so if there are N sites on the lattice, it follows that there are $N - 1$ excited states.

Let's calculate the change in free energy that occurs because of this excitation. First, recall that the free energy is $F = E - TS$. In this case, the excitation increases the energy E by the amount $2J$. In addition, it occurs in $(N - 1)$ locations, and hence it leads to an increase in entropy S of $k_B \ln(N - 1)$. As a result, the change in free energy is

$$\Delta F = 2J - k_B T \ln(N - 1).$$

Notice that the energy term increases the free energy, while the entropy term decreases it.

Now, in general, a thermodynamic system seeks to lower its free energy. In this case, we can see that for any finite temperature, T, the entropy term will dominate in the limit of $N \to \infty$, and hence the free energy is lowered as a result of the excitation. This means that the excitation is favored at finite T and occurs freely throughout the system. But note that the excitation flips a finite fraction of

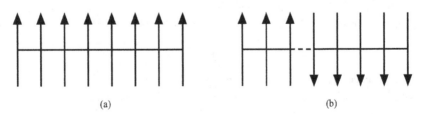

(a)   (b)

Figure 5.9 (a) One of the ground states of the 1-D chain. (b) One of the first-excited states.

the spins, destroying any possible ordered phase; that is, there is no magnetization for any finite T. As a result, there is no finite-temperature phase transition.

This same argument can be applied to the $N \times N$ two-dimensional lattice, but with the opposite conclusion. Once again, the first excitation, as shown in Figure 5.7 (b), will proliferate throughout the system at any finite temperature. That is the same. What is different, however, is that this excitation is just a single spin flipped out of an infinite lattice – and this has a vanishingly small effect on the magnetization. To flip half of the lattice and destroy the magnetization, as in one dimension, would require breaking at least N bonds – but this would make the energy cost significantly greater than the entropy gain. Using arguments along these lines, Peierls was able to show rigorously that the magnetization persists to a finite temperature in the two-dimensional lattice. The actual value of that finite temperature was obtained several years later using duality, which we cover in Section 5.4.

5.3 High-Temperature Series Expansions

To expand the reduced free energy for a two-dimensional lattice of N sites in the limit of high temperature T means that we must first do an expansion of the partition function Z_N for small K. We will then take the log of Z_N, divide by N, and take the limit $N \to \infty$.

To begin, let's recall that the partition function is

$$Z_N = \sum_{\{s_i\}} e^{K\sum_{\langle i,j \rangle} s_i s_j} = \sum_{\{s_i\}} \prod_{\langle i,j \rangle} e^{K s_i s_j}. \tag{5.8}$$

Notice that in this expression we have rewritten the exponential term, which originally is raised to the power of the *sum* over all nearest neighbors, as a *product* of exponentials, each raised to the power of one nearest-neighbor pair. This will be most helpful as we move forward.

With this in mind, let's consider the Boltzmann factor for one of the nearest-neighbor pairs of spins:

$$e^{K s_i s_j}.$$

The straightforward thing to do is to simply expand the exponential for small-K, which yields the following:

$$e^{K s_i s_j} = 1 + K(s_i s_j) + \frac{1}{2!}K^2(s_i s_j)^2 + \frac{1}{3!}K^3(s_i s_j)^3 + \frac{1}{4!}K^4(s_i s_j)^4 + \dots. \tag{5.9}$$

We can obtain a much more useful result, however, if we take advantage of one of the basic properties of Ising spins – namely, that they take on only two values: +1 and –1. As a result, it follows that $(s_i s_j)^2 = s_i^2 s_j^2 = 1$ and $(s_i s_j)^3 = (s_i s_j)^2 (s_i s_j) = (s_i s_j)$. We can generalize these results as follows:

$$(s_i s_j)^n = 1 \qquad \text{for even } n$$
$$(s_i s_j)^n = (s_i s_j) \qquad \text{for odd } n.$$

Thus, all the powers of $(s_i s_j)$ in Equation 5.9 reduce to either 1 or $(s_i s_j)$.

Applying this simplification, we can rearrange the expansion of $e^{K s_i s_j}$ as follows:

$$e^{K s_i s_j} = \left[1 + \frac{1}{2!}K^2 + \frac{1}{4!}K^4 + ... \right] + \left[K + \frac{1}{3!}K^3 + \frac{1}{5!}K^5 + ... \right](s_i s_j). \quad (5.10)$$

Now, the two expansions in square brackets in Equation 5.10 may or may not look familiar, but here's an easy way to evaluate them. We have just two independent expansions, call them a and b, and just two possible values for $(s_i s_j)$, thus we can write

$$e^{K s_i s_j} = a + b(s_i s_j).$$

We can evaluate the two unknowns, a and b, as follows:

$$e^K = a + b$$
$$e^{-K} = a - b.$$

Rearranging and solving yields

$$a = \cosh K$$
$$b = \sinh K.$$

These are, in fact, the expansions given in Equation 5.10. Thus, we can now write

$$e^{K s_i s_j} = \cosh K + \sinh K(s_i s_j) = \cosh K[1 + v(s_i s_j)]. \quad (5.11)$$

The term v in this last expression is the hyperbolic tangent:

$$v = \tanh K.$$

Notice that v is small for small K, and hence from this point on our expansion parameter will be v rather than K.

Evaluating the Partition Function: Graphical Analysis

Returning to the partition function, we can now write it as follows:

$$Z_N = \sum_{\{s_i\}} \prod_{\langle i,j \rangle} e^{Ks_is_j} = (\cosh K)^{2N} \sum_{\{s_i\}} \prod_{\langle i,j \rangle} [1 + v(s_is_j)]$$

$$= (\cosh K)^{2N} \sum_{\{s_i\}} \{[1 + v(s_is_j)][1 + v(s_ks_l)]...\}.$$

$$(5.12)$$

The $(\cosh K)^{2N}$ term in front of the summation simply represents one factor of $(\cosh K)$ for each of the $2N$ nearest-neighbor pairs. The product over nearest neighbors, on the other hand, is best evaluated in terms of *graphs* on the lattice.

To see how this can be done, notice that each term in the product has either a 1 or a v for each nearest-neighbor pair. Graphically, we associate each v with a line, or *bond*, connecting the corresponding nearest neighbors. If a nearest-neighbor pair has the term 1 instead, we represent that by the absence of a bond.

For example, in Figure 5.10 (a) we show the "empty" graph (no bonds), which has a 1 associated with each nearest neighbor pair. The contribution this graph makes to the partition function is

$$(\cosh K)^{2N} \sum_{\{s_i\}} 1 = 2^N (\cosh K)^{2N}.$$

Notice that the sum over all configurations of N spins yields 2^N, as expected.

Figure 5.10 (b) shows the graph with v for one nearest neighbor pair, and 1 for all the others. The contribution corresponding to this graph is

$$(\cosh K)^{2N} \sum_{\{s_i\}} v(s_is_j) = 0.$$

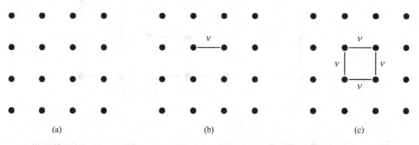

(a)   (b)   (c)

Figure 5.10 Various high-temperature graphs on the 2-D square lattice.

The fact that this sum is equal to zero follows because $\sum_{\{s_i\}} s_i = 0$ for any spin s_i.

The simplest nontrivial graph that has a nonzero contribution is an elementary square of four bonds, as shown in Figure 5.10 (c). The contribution to the partition function in this case is

$$(\cosh K)^{2N} \sum_{\{s_i\}} v^4 (s_i s_j)(s_j s_k)(s_k s_l)(s_l s_i) = 2^N (\cosh K)^{2N} v^4.$$

This is the same contribution as for the empty graph, except for the inclusion of the term v^4, which represents the four bonds.

Looking at these results, it becomes clear that the only way a graph can give a nonzero contribution is if it has an even number of bonds (0, 2, 4) associated with each site. Any site with an odd number of bonds will have s_i raised to an odd power, and that will sum to zero.

The contributing graphs to order v^4 and v^6 are shown in Figure 5.11. Notice that there is one graph per site (N) for order v^4. Similarly, there are two graphs per site ($2N$) for order v^6, one vertical and one horizontal.

Combining these results, we can write the partition function to order v^6 as follows:

$$Z_N = 2^N (\cosh K)^{2N} (1 + Nv^4 + 2Nv^6 + \ldots).$$

The corresponding reduced free energy per site is

$$f(K) = \lim_{N \to \infty} \left[\frac{1}{N} \ln Z_N \right] = \ln 2 + 2 \ln(\cosh K) + v^4 + 2v^6 + \ldots.$$

The specific heat is easily evaluated with $c = K^2 \partial^2 f / \partial K^2$, and the results for the three nontrivial terms just given are plotted on the small-K side of

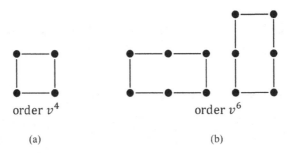

order v^4 order v^6

(a)   (b)

Figure 5.11 High-temperature contributing graphs. (a) The single contributing graph per site to order v^4. (b) The two contributing graphs per site to order v^6.

Figure 5.8. We see again that the approximation becomes better with each additional term in the series, and that the singularity will only be seen in the limit of an infinite number of terms.

Graphs in One Dimension

The graphical analysis of this section can be applied to any dimension – one, two, three, four, and so on – as well as to any type of lattice structure – square, triangular, honeycomb, cubic, face-centered cubic, and so on. The application to one dimension is particularly enlightening, since it leads to the exact solution.

Let's start with a one-dimensional chain (free boundary conditions) of N spins. The lattice is shown in Figures 5.12 (a) and (b). The empty graph is shown in Figure 5.12 (a), and the contribution this graph makes to the partition function is

$$Z_{N,\text{chain}} = 2^N (\cosh K)^{N-1}. \tag{5.13}$$

There are no other contributing graphs. As an example, consider the graph in Figure 5.12 (b). This graph, and all others you could draw, will have sites with an odd number of bonds – and hence zero contribution. Thus, the exact partition function for this system is the result given in Equation 5.13, in agreement with Equation 4.2. Quite simple.

Next, let's consider a one-dimensional Ising system with periodic boundary conditions; that is, a ring lattice, as in Figures 5.12 (c) and (d). The analysis here is almost as simple as it is for the chain. Again, we have the empty graph, as in Figure 5.12 (c). This has the contribution given in Equation 5.13, except that $N-1$ (the number of bonds in the chain lattice) is replaced by N (the number of bonds in the ring lattice) in the exponent of the ($\cosh K$) term. We have another

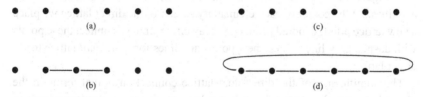

Figure 5.12 (a) The one-dimensional chain lattice (free boundary conditions) with no bonds. (b) Any graph with a finite number of bonds on the 1-D chain, like the one shown here, will have zero contribution. (c) The one-dimensional ring lattice, with periodic boundary conditions. This is the empty graph. (d) The full graph for the ring lattice.

contributing graph, however – the full graph; that is, the graph with N bonds connecting all N sites, as in Figure 5.12 (d). The contribution from this graph is $2^N (\cosh K)^N v^N$. It follows that the complete partition function for the ring is

$$Z_{N,\text{ring}} = 2^N (\cosh K)^N (1 + v^N). \qquad (5.14)$$

This result agrees with Equation 4.34, as expected.

5.4 Duality and the Location of the Critical Point

It's useful to compare the high- and low- temperature series expansions for the two-dimensional square lattice. Specifically, the partition functions for high temperature (HT) and low temperature (LT) are as follows:

$$Z_{N,\text{HT}} = 2^N (\cosh K)^{2N} [1 + Nv^4 + 2Nv^6 + \ldots]$$

$$Z_{N,\text{LT}} = 2e^{2NK} [1 + N(e^{-2K})^4 + 2N(e^{-2K})^6 + \ldots].$$

In each case, the partition function starts off with prefactors that are analytic; that is, no singularities will arise from $\cosh K$ or e^K raised to various powers. It follows that any singularities associated with a phase transition must occur in the infinite series expansions.

Now, when we look carefully at these series, we notice that they have *exactly* the same structure:

$$1 + N(\text{expansion parameter})^4 + 2N(\text{expansion parameter})^6 + \ldots.$$

In the HT case the expansion parameter is $v = \tanh K$, and in the LT case it is e^{-2K}. Other than that, the series are the same – at least to this order. It turns out, in fact, that the series expansions continue to be the same *to all orders*, as can be seen by considering what is referred to as the *dual* lattice.

In Figure 5.13 we show the original, or *direct*, lattice with solid black dots and lines. At the center of each elementary square of the direct lattice we place a new lattice point, denoted by an open gray circle, and we connect these points with dashed gray lines. These new points and lines form the *dual* lattice to the direct lattice.

The significance of the direct/dual lattice connection is that terms in the HT series expansion on the direct lattice have a one-to-one correspondence to terms in the LT series expansion on the dual lattice, and vice versa. To see how this works, consider Figure 5.14. On the left we see the smallest contributing graph in the HT expansion – an elementary square of bonds.

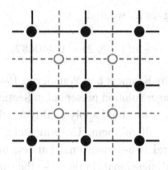

Figure 5.13 The solid lines and dots are the direct lattice; the dashed lines and open dots are the dual lattice.

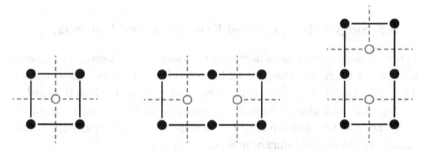

Figure 5.14 High-temperature contributing graphs on the direct lattice correspond to low-temperature contributing graphs on the dual lattice.

If we imagine flipping the spin on the dual lattice in the middle of this square, we have the first excitation from LT – a flipped spin with four broken bonds. Similarly, the other two graphs in Figure 5.14 show the next order (v^6) graphs from HT, and again, if we flip the dual spins contained within these graphs we obtain the second-order excitations from LT – namely, two nearest-neighbor spins flipped with six broken bonds. This one-to-one connection continues to all orders.

We can see now that if one partition function series has a singularity at a certain value of its expansion parameter, then the other series must have a singularity at that same value of the expansion parameter. We refer to this as a *self-dual point*. Written out mathematically, what we're saying is that if K_c is the critical point, then the self-dual condition is

$$e^{-2K_c} = v_c = \tanh K_c.$$

Rearranging and solving for K_c we find

$$K_c = \frac{1}{2}\ln(1 + \sqrt{2}) = 0.440687\ldots\quad\quad (5.15)$$

This is the value found by Kramers and Wannier in 1941, and it is indeed the critical point in the Onsager solution presented in Section 5.1.

Now, it would be nice if we could apply duality to other lattices and obtain exact critical points for all of them. Unfortunately, this is not the case. Duality works for several two-dimensional lattices, but it doesn't apply to any three- or higher-dimensional lattices – the one-to-one connection between direct and dual lattices simply breaks down in dimensions higher than two.

5.5 Singularities, Critical Exponents, and Universality

In the previous section, we determined the value of the expansion parameter that gives a singularity (nonanalytic behavior) in the partition function, Z, of a 2-D Ising model. It follows that the reduced free energy, f, which is related to the log of Z, will also be singular. In addition, thermodynamics quantities, which are given by derivatives of f, will display singularities as well. In this section we discuss the nature of these singularities.

To begin, we define a *reduced temperature*, t, that gives a dimensionless measure of the difference between the temperature T and the critical temperature T_c:

$$t = \frac{|T - T_c|}{T_c}.$$

Notice that $t \to 0$ as $T \to T_c$. It turns out that Ising model singularities have a power law dependence on the variable t near the critical point.

Critical Exponents

If a thermodynamic quantity Q vanishes at the critical point, we write it in the limit of small t as follows:

$$Q \sim t^E.$$

In this expression, E is referred to as the *critical exponent*. It follows that $Q \to 0$ as $t \to 0$ for positive E. Similarly, if Q diverges to infinity at the critical point, we will write it a little differently:

$$Q \sim t^{-E}.$$

Notice that we've included a minus sign in the exponent this time, so that if E is positive, then $Q \to \infty$ as $t \to 0$. In general, we always define critical exponents to have positive values by including negative signs where needed.

The most common critical exponents are those related to the specific heat, c (exponent $= \alpha$), the magnetization, m (exponent $= \beta$), and the magnetic susceptibility, χ (exponent $= \gamma$). Near the critical point, we can write these thermodynamic functions as follows:

$$c \sim t^{-\alpha}$$

$$m \sim t^{\beta}$$

$$\chi \sim t^{-\gamma}.$$

Notice that we've included minus signs for the quantities (c and χ) that diverge at the critical point.

Now, the critical exponent α is a bit of a special case, so we'll start with β and come back to α later. Referring to Figure 5.6, we see that m vanishes at the critical point. The expression in Equation 5.6 immediately indicates that the exponent in this case is $\beta = 1/8$ (we shall give the details of this conclusion in the next chapter). In three dimensions, the corresponding value is $\beta = 0.326$ .... Table 5.1 collects these results, as well as all the other critical exponents discussed in this section.

Next, we consider χ, which is a measure of the fluctuation of the magnetization. This quantity diverges to infinity at the critical point. The exponent describing this divergence in two dimensions is $\gamma = 7/4$. In three dimensions, the approximate value of the exponent is $\gamma = 1.23$ ....

Finally, we return to the specific heat and its exponent, α. The value of this exponent in two dimensions is $\alpha = 0$. What does this mean? Well, the most obvious answer is that the specific heat approaches a constant value at the critical point, say with one constant value above the critical point and another constant value below it. This means a *discontinuity* in the specific heat, which is certainly a singularity. A discontinuity in the specific heat is seen in four dimensions, and in mean field theory, as indicated in Table 5.1.

There is another possibility, however. Consider the following definition for the natural log:

$$\lim_{\alpha \to 0} \frac{t^{\alpha} - 1}{\alpha} = \ln t.$$

Table 5.1 *Critical exponents for various dimensions. The results for 2-D and 4-D are exact; those for 3-D are approximations obtained from series expansion analyses.*

	α	β	γ
2-D	0 (log)	1/8	7/4
3-D	0.110	0.326	1.23
4-D (mean field)	0 (discontinuous)	1/2	1

This shows that $\alpha = 0$ can also indicate a logarithmic divergence at the critical point; that is,

$$c \sim \ln t.$$

This, in fact, is the case for the two-dimensional Ising model. Thus, the specific heat in two dimensions diverges, though it does so more slowly than any positive power of t. In three dimensions, it is found that $\alpha = 0.110....$

Universality

The results shown in Table 5.1 illustrate one of the key properties of critical exponents; namely, the concept of *universality classes*. For example, we show the values for the exponents in 2-D in the top row of Table 5.1, though we don't specify which of the two-dimensional lattices we have in mind. That's because it doesn't matter – the critical exponents are the same on the square, triangular, hexagonal, and other 2-D lattices. The thermodynamic functions and the location of the critical points will vary from lattice to lattice, but close to the critical point each system behaves the same.

Thus, we can say that the top row in Table 5.1 represents the *2-D Ising universality class*. In general, universality classes are determined by large-scale, global features of a system – like the dimension and the type of spin – and are unaffected by small-scale, local features – like the number of nearest neighbors or the detailed interactions. So, for example, adding a second-neighbor interaction in two dimensions will change the location of the critical point, but it won't change the exponents.

The next row in Table 5.1 gives the critical exponents for the *3-D Ising universality class*. Notice that these exponents differ from the corresponding two-dimensional values, but again they don't depend on details like the specific structure of the 3-D lattice. All of these conclusions are the result of decades of hard work analyzing series expansions, and applying other techniques as well.

The concept of universality classes was discovered as a result of the hard work done on these systems dating back to the 1960s.

The bottom row in Table 5.1 shows the results for a four-dimensional lattice. Now, of course, there are no four-dimensional crystals in nature. From a theoretical perspective, however, there's no problem considering a four- or higher-dimensional lattice – in fact, one can gain valuable insight in this way. For example, we find that the critical exponents are the same for *all* dimensions equal to or greater than 4; thus, 4-D is a sort of boundary – below 4-D critical exponents vary with dimension, but above it they remain the same as the dimension is changed. It turns out that fluctuations are much less important in higher dimensions, and hence all higher dimensions behave the same near the critical point.

In addition, we shall see in the next chapter that the exponents in 4-D correspond exactly to the exponents found in the approximation methods referred to as mean-field theories. This makes sense, because mean-field theories ignore fluctuations – which is a good approximation in higher dimensions. We shall pursue these concepts in greater detail in the next chapter.

5.6 Problems

5.1 Give the high-temperature expansion of the reduced free energy per site for the ladder lattice to order v^6, where $v = \tanh K$.

5.2 Give the low-temperature expansion of the reduced free energy per site for the simple cubic lattice to order $(e^{-2K})^{10}$.

5.3 Give the low-temperature expansion of the reduced free energy per site for the triangular lattice to order $(e^{-2K})^{10}$.

5.4 Give the low-temperature expansion of the reduced free energy per site for the hexagonal lattice to order $(e^{-2K})^4$.

6

The Mean-Field Approach

Exact solutions for infinite Ising systems are rare, specific in terms of the interactions allowed, and limited to one and two dimensions. To study a wider range of models we must resort to various approximation techniques. One of the simplest and most comprehensive of these is the mean-field approximation, the subject of this chapter. Some versions of this approximation rely on a self-consistent requirement, and in this respect the mean-field method for the Ising model is similar to a number of other self-consistent approximation methods in physics, including the Hartree–Fock approximation for atomic and molecular orbitals, the BCS theory of superconductivity, and the relaxation method for determining electric potentials. We will also introduce a somewhat different mean-field approach, the Landau–Ginzburg approximation, which is based on a series expansion of the free energy. One of the drawbacks of *all* the mean-field theories, however, is that they predict the same mean-field critical exponents, which, unfortunately, are at odds with the results of exact solutions and experiments.

6.1 Mean-Field Theory for a Single Spin

In the simplest version of the mean-field approach for the Ising model we focus on a single spin on the lattice, which we label s_0. The underlying premise is that each spin on an infinite lattice is the same as all the other spins, and hence any spin is representative of the behavior of the entire lattice. The other spins are to be replaced in our calculations with their average, or *mean*, value – hence the name of the method.

To keep things simple, and easy to illustrate, we consider the two-dimensional square lattice with a nearest-neighbor interaction energy J. The Hamiltonian for the system is

$$H = -J \sum_{\langle ij \rangle} s_i s_j.$$

As in previous chapters, the summation is over all nearest-neighbor pairs of spins on the lattice. If J is positive, we see that the energy is lowest for spins that are aligned; that is, for spins that have the same sign. Thus $J > 0$ represents a ferromagnetic interaction, and this will be the focus of our attention in what follows.

The situation for this calculation is shown in Figure 6.1. On the left-hand side we show s_0 surrounded by the four nearest-neighbor spins of the square lattice; specifically, s_1, s_2, s_3, s_4. The first step is to replace the nearest-neighbor spins with their average values. Since all spins on the lattice are equivalent to one another, each of the average values is simply equal to the magnetization, m. That is,

$$\langle s_1 \rangle = \langle s_2 \rangle = \langle s_3 \rangle = \langle s_4 \rangle = m.$$

The situation following this replacement is shown on the right-hand side of Figure 6.1.

Notice that each nearest-neighbor bond has the strength J. The number of nearest neighbors is designated as q, which is referred to as the *coordination number*. For the square lattice, shown in Figure 6.1, the coordination number is $q = 4$. For the triangular lattice the coordination number is $q = 6$, and for the simple cubic lattice it is also $q = 6$. It follows that the triangular and simple cubic lattice are precisely the same in this approximation. This is a severe oversimplification of the mean-field approach, given that dimensionality actually has a significant impact on the behavior of the Ising model. We will revisit this topic later in the section.

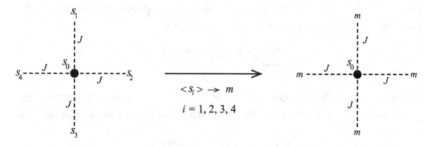

Figure 6.1 The basic mean-field calculation for a single spin s_0. The q spins surrounding s_0 are replaced with their average value, m. They interact with s_0 through bonds of strength J.

The Hamiltonian for the single spin s_0 can now be written as follows:

$$H_0 = -qJms_0.$$

Similarly, the reduced Hamiltonian, which is useful in calculating probabilities, among other things, is

$$-\beta H_0 = q\frac{J}{k_B T}ms_0.$$

In the past we have introduced the dimensionless coupling K, defined as $K = J/k_B T$, but in this case we will continue to use the temperature T as our basic variable. The reason for this is that the magnetization is nonzero over a finite range of temperatures, from 0 to the critical temperature, T_c, whereas, on the other hand, the magnetization is nonzero for values of the coupling that range from the critical value, K_c, to infinity. The finite range of values for T makes plots in terms of this variable much more convenient.

Self-Consistency

We come now to the heart of the mean-field method – the self-consistent requirement. We've assumed that the magnetization of all the spins surrounding s_0 are equal to m. The bonds connecting these spins to s_0 tend to align it in the same direction as the other spins on the lattice, assuming, as we do, that J is positive. This means that s_0 has a magnetization – and since it is like any other spin on the lattice, its magnetization must also be equal to m. This condition, namely that $<s_0> = m$, makes the approximation self-consistent.

To calculate the average value of s_0, we need the probability that s_0 is equal to $+1$ or -1. Probabilities are proportional to the Boltzmann factor, $e^{-\beta H_0}$, and the factor that converts the proportionality to a specific probability is the partition function. The partition function for s_0 is

$$Z_0 = \sum_{\{s_0\}} e^{-\beta H_0} = e^{q\frac{J}{k_B T}m} + e^{-q\frac{J}{k_B T}m}.$$

The first term in the sum corresponds to $s_0 = +1$, and the second term corresponds to $s_0 = -1$. It follows that the probability that the spin s_0 is equal to $+1$ is

$$P[s_0 = +1] = \frac{1}{Z_0}e^{q\frac{J}{k_B T}m}.$$

Similarly, the probability that s_0 is equal to -1 is

$$P[s_0 = -1] = \frac{1}{Z_0} e^{-q\frac{J}{k_B T}m}.$$

Therefore, the average value of s_0 is given by the following:

$$\langle s_0 \rangle = \frac{1}{Z_0} \sum_{\{s_0\}} s_0 e^{-\beta H_0} = \frac{e^{q\frac{J}{k_B T}m} - e^{-q\frac{J}{k_B T}m}}{e^{q\frac{J}{k_B T}m} + e^{-q\frac{J}{k_B T}m}} = \tanh\left[q\frac{J}{k_B T}m\right].$$

Now, applying the self-consistency requirement, $\langle s_0 \rangle = m$, we obtain the following relation:

$$m = \tanh\left[q\frac{J}{k_B T}m\right]. \tag{6.1}$$

The solution to this equation for a given temperature T is the magnetization m that characterizes every spin on the lattice.

General Behavior and Critical Temperature

Equation 6.1 cannot be solved analytically – and for good reason, since the behavior of m is inherently nonanalytic. Instead, we can find numerical solutions for m by graphing both sides of Equation 6.1 and looking for intersection points, which is where the two sides are equal. This is illustrated in Figure 6.2. In this plot, the straight line passing through the origin is the left-hand side of Equation 6.1; namely, $y = m$. The three sigmoidal-shaped curves represent the tanh function on the right-hand side of Equation 6.1 for a variety of temperatures. The intersection of $y = m$ with one of the tanh curves gives a point – a value of m – where Equation 6.1 is satisfied for a specific temperature.

The first notable feature of Figure 6.2 is that there is always a solution at $m = 0$, no matter what the temperature. This is easily verified by substituting $m = 0$ into Equation 6.1. It is also clear from physical considerations – after all, if the bulk of the spins on the lattice have $m = 0$, then they will not induce a magnetization in s_0, and hence its average value will be zero as well. At high temperatures $m = 0$ is the *only* solution, and it corresponds to a completely disordered lattice.

At low temperatures two additional solutions appear, one positive and one negative, as shown by the upper curve on the right-hand side of Figure 6.2. These two solutions are symmetric about $m = 0$, and correspond to the fact that the zero-field Ising model is unchanged by flipping all the spins. In addition, the finite-m solutions have a lower free energy than the $m = 0$ solutions at any given temperature, as we shall see later in this chapter, and hence they are the solutions

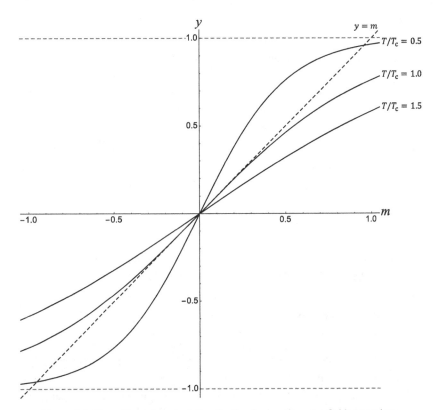

Figure 6.2 The self-consistent relation for the single-spin mean-field approximation. The straight line passing through the origin is the left-hand side of
Equation 6.1; that is, $y = m$. The three curves passing through the origin are the
right-hand side of Equation 6.1 for three different values of the temperature.

that characterize the physical system. As the temperature is decreased toward
zero, the finite-m solutions approach $+1$ and -1, indicating that the lattice is
completely ordered with all spins up or all spins down. We will focus on the
positive-m solutions.

Separating the regions of finite-m and zero-m solutions is the critical temperature T_c, which corresponds to the middle curve in Figure 6.2. At this
temperature, the initial slope of the tanh curve is equal to 1, the same as the
slope of the line $y = m$. To find an expression for the critical temperature, we
note that the series expansion for $\tanh(ax)$ for small x and constant a is

$$\tanh(ax) = ax + \mathrm{O}(x^3). \tag{6.2}$$

Applying this expansion to the right-hand side of Equation 6.1 yields

$$\tanh\left[q\frac{J}{k_\mathrm{B}T}m\right] \approx q\frac{J}{k_\mathrm{B}T}m.$$

Thus, the initial slope of the tanh function is $q\frac{J}{k_\mathrm{B}T}$, and setting this equal to 1 yields the critical temperature T_c. To be specific, we find

$$q\frac{J}{k_\mathrm{B}T_\mathrm{c}} = 1. \tag{6.3}$$

Rearranging gives

$$T_\mathrm{c} = q\frac{J}{k_\mathrm{B}}. \tag{6.4}$$

Thus, the key factor in determining the critical temperature, and the factor that distinguishes one lattice from another, is the coordination number, q.

For example, on the 2-D square lattice, where $q = 4$, we have $T_\mathrm{c} = (4)\frac{J}{k_\mathrm{B}}$. For comparison, the exact result is $T_\mathrm{c} = (2.27...)\frac{J}{k_\mathrm{B}}$. On the triangular lattice, with $q = 6$, the mean-field critical temperature is $T_\mathrm{c} = (6)\frac{J}{k_\mathrm{B}}$. This is to be compared with the exact result on the triangular lattice, which is $T_\mathrm{c} = (3.64)\frac{J}{k_\mathrm{B}}$. For the simple cubic lattice, the mean-field critical temperature is the same as it is for the triangular lattice, since these lattices have the same coordination number. The best estimate for the critical temperature on the simple cubic lattice is $T_\mathrm{c} = (4.51)\frac{J}{k_\mathrm{B}}$, which is considerably closer to the mean-field result. We collect these results, along with other critical properties, later in this section. In general, mean-field results are better the higher the dimension, where fluctuations about the mean value – which are ignored in this approach – are less important.

Now that we've determined the critical temperature, it's useful to rewrite the self-consistency relation in terms of the dimensionless temperature variable T/T_c. Starting with Equation 6.1, let's multiply and divide by T_c in the argument of the tanh to find the following:

$$m = \tanh\left[q\frac{J}{k_\mathrm{B}T}m\right] = \tanh\left[q\frac{J}{k_\mathrm{B}T_\mathrm{c}}\left(\frac{T_\mathrm{c}}{T}\right)m\right].$$

Substituting the expression for the critical temperature, Equation 6.3, this simplifies to

$$m = \tanh\left[\left(\frac{T}{T_\mathrm{c}}\right)^{-1}m\right]. \tag{6.5}$$

The three curves in Figure 6.2, from top to bottom, are plots of the right-hand side of this equation for the values $T/T_c = 0.5$, 1.0, and 1.5, respectively.

Before continuing, we should mention a particularly severe drawback of the mean-field approximation – namely, that it predicts a finite critical temperature for the 1-D Ising model. For example, the 1-D chain lattice has a coordination number given by $q = 2$, which implies a critical temperature of $T_c = (2)\frac{J}{k_B}$. In fact, as we know from Chapter 4, there is *no* phase transition in one dimension, and hence the actual critical temperature is *zero*. The mean-field method predicts a phase transition in 1-D because, as we've stated, it ignores fluctuations, and fluctuations, as we saw in Section 5.2, destroy the ordered state in one dimension at any finite temperature.

Numerical Solutions and Cobwebbing

Finding numerical solutions to Equation 6.5 is fairly easy to do. Of course, one can always make a guess, and then refine the guess to get closer and closer to the desired solution. However, since the tanh curve on the right-hand side of Equation 6.5 crosses the line $y = m$ with a slope less than 1, it follows that successive iterations of the tanh function will automatically converge to the desired solution. No guessing is required. It's as if the equation solves itself.

To see how this works, we can think of Equation 6.5 as a relation that iterates from one approximation for the magnetization m to the next. Thus, if m_n is the current approximation to the solution, the right-hand side of the equation gives the next approximation, m_{n+1}, which is closer to the true solution. Specifically,

$$m_{n+1} = \tanh\left[\left(\frac{T}{T_c}\right)^{-1} m_n\right].$$

Iterating this relation gives the solution to any desired accuracy. The ultimate result – the solution we seek – can then be seen as a *fixed point* of this iteration, which we designate as m^* and define as follows:

$$m^* = \tanh\left[\left(\frac{T}{T_c}\right)^{-1} m^*\right]. \tag{6.6}$$

For each value of $T/T_c < 1$ there is a corresponding finite, positive value of m^*.

For example, suppose we would like to find the numerical value of m^* for the case $T/T_c = 0.5$. This corresponds to finding the intersection point of the tanh curve and the $y = m$ straight line in Figure 6.3. For our initial guess, which we will call m_0, let's take $m_0 = 0.5$ – we could make a much better guess by

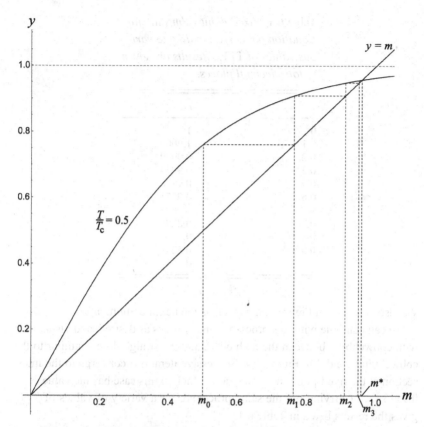

Figure 6.3 Finding the solution to Equation 6.5 for the case $T/T_c = 0.5$. The successive iterations of the equation can be viewed as cobwebbing between the straight line (the left-hand side of Equation 6.5) and the tanh curve (the right-hand side of Equation 6.5).

referring to Figure 6.3, but this crude choice is useful for illustrative purposes. The next iteration, m_1, is given by

$$m_1 = \tanh\left[\left(\frac{T}{T_c}\right)^{-1} m_0\right] = \tanh[2(0.5)] = 0.761594\ldots.$$

We show both of these values of m on the horizontal axis in Figure 6.3. The next value of the magnetization, m_2, is given by

$$m_2 = \tanh\left[\left(\frac{T}{T_c}\right)^{-1} m_1\right] = \tanh[2(0.761594\ldots)] = 0.909252\ldots.$$

Table 6.1 *Fixed point values* m* *for Equation 6.6 corresponding to various values of* T/T_c. *Results are given to four decimal places.*

T/T_c	m^*
0	1
0.1	1.0000
0.2	0.9999
0.3	0.9974
0.4	0.9856
0.5	0.9575
0.6	0.9073
0.7	0.8286
0.8	0.7104
0.9	0.5254
1	0

We also show m_2 in Figure 6.3, as well as the next iteration, m_3.

We can continue with this process as many times as desired, and visualize it with cobwebbing between the tanh curve and the straight line – similar to the cobwebbing we did in Figure 3.4. Successive iterations converge to the intersection at the fixed-point magnetization, which in this case has the value $m^* = 0.9575 \ldots$. Carrying out the same procedure for a variety of values of T/T_c gives the results listed in Table 6.1.

The data points in Table 6.1 are plotted in Figure 6.4, along with the overall solution, which is represented by the solid curve. Notice that the magnetization, $m = m^*$, saturates to $m = 1$ as the temperature goes to zero, and goes to zero, $m = 0$, at $T = T_c$. The magnetization remains zero for all higher temperatures.

The fact that the magnetization is zero for all temperatures above T_c, but is nonzero below that temperature, means that the magnetization is a nonanalytic function. There is no analytic function, in fact, that can be zero over a finite range of temperatures and then abruptly become nonzero – there must be a singularity at the switchover point. This is precisely why there is no analytic solution to Equation 6.5. In addition, as we will show later in this section, the singularity at the critical temperature T_c is a power-law singularity, characterized by a *critical exponent*. While this type of singularity is also predicted by Onsager's exact solution for the 2-D Ising model, the mean-field prediction for the *value* of the *exponent* is different from the exact result, and is also different from the value seen in experiments.

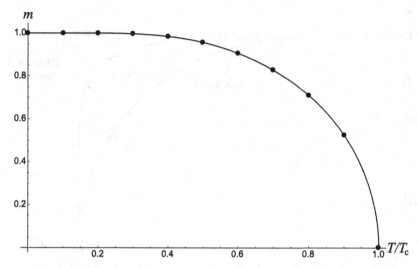

Figure 6.4 The mean-field magnetization solutions to Equation 6.6, $m = m^*$, as a function of T/T_c.

Comparing Magnetizations

One might wonder whether the magnetization curve shown in Figure 6.4 gives results for the square lattice, the triangular lattice, or perhaps the simple cubic or face-centered cubic lattice. In fact, it represents all of them.

To see this, note that the temperature scale in Figure 6.4 is normalized to the critical temperature; that is, we plot m versus T/T_c. Now, each lattice has its own specific value of T_c, which in general are quite different. Once we normalize the temperature to this value of T_c, however, the magnetization for each lattice vanishes at the same point, $T/T_c = 1$. All the other points on the curve are the same for all lattices as well. This can be seen by referring to Equation 6.5, where we see that m depends only on the variable T/T_c; all of the dependence on the lattice type has been encapsulated into the value of T_c. In this respect, we can think of the magnetization in Figure 6.4 as the *universal*, mean-field magnetization for a *single-spin* cluster on any lattice.

Let's turn now to a comparison between our mean-field magnetization and the exact results for 2-D and 3-D lattices. In Figure 6.5, the mean-field magnetization is the lowest curve (solid). The uppermost curve (dashed) is the exact magnetization for the 2-D square lattice, as given by Equation 5.6. It differs significantly from the mean-field result. This isn't particularly surprising, of course, given that the comparison between 1-D and mean-field isn't

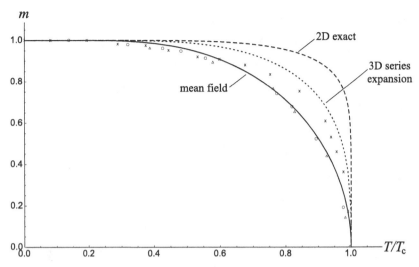

Figure 6.5 A comparison of the mean-field magnetization (solid curve), exact 2-D square lattice magnetization (dashed curve), and exact 3-D simple cubic magnetization (dotted curve). The experimental data points are for iron (X), nickel (O), and cobalt (Δ).

even close. Better agreement is found with the results for the 3-D simple cubic lattice, shown by the middle curve (dotted). These results are from exact series expansions, and can be considered to be exact to within the precision of the graph. Notice that as the dimension increases, the results of mean-field theory generally tend to improve – a direct reflection of the fact that fluctuations have less impact the greater the dimension.

Finally, three different series of experimental data points are shown for comparison as well. The points shown with (X) are from experiments on iron, those with (O) are from nickel, and those with (Δ) are from cobalt. The agreement between the mean-field theory and experiment is quite reasonable, especially given the simplicity of the single-spin, mean-field approximation.

Magnetization Critical Exponent: Single-Spin Mean Field Theory

As has been mentioned, the mean-field magnetization has a power-law singularity as the critical temperature is approached from below. We will now derive this result, as well as the precise value of the exponent in the power-law dependence. The exponent is of particular interest in critical phenomena because, as it turns out, it has the same (universal) value over a wide range of

Ising models, depending mostly on the dimension of the system, but not at all on many other details.

To explore the behavior near the critical point, we expand the tanh function on the right-hand side of Equation 6.5 for small values of the magnetization m. We expanded this function in Equation 6.2 to find the critical temperature, but this time we carry the expansion to one higher order in small quantities. The result is

$$\tanh(ax) = ax - \frac{1}{3}(ax)^3 + O(x^5). \tag{6.7}$$

Applying this to Equation 6.5, we find

$$m = \tanh\left(\left(\frac{T}{T_c}\right)^{-1} m\right) \cong \left(\frac{T}{T_c}\right)^{-1} m - \frac{1}{3}\left(\frac{T}{T_c}\right)^{-3} m^3.$$

Rearranging, and canceling one power of m, yields

$$m^2 = 3\left(\frac{T}{T_c}\right)^3 \left(\left(\frac{T}{T_c}\right)^{-1} - 1\right) = 3\left(\frac{T}{T_c}\right)^2 \left(1 - \left(\frac{T}{T_c}\right)\right). \tag{6.8}$$

At this point it's useful to introduce the *normalized reduced temperature, t,* defined as follows:

$$t = 1 - \left(\frac{T}{T_c}\right).$$

Notice that t vanishes linearly as T approaches T_c from below. Substituting this definition into Equation 6.8 yields

$$m^2 = 3\left(\frac{T}{T_c}\right)^2 t.$$

Taking the square root of both sides gives the temperature dependence of m near T_c:

$$m = \sqrt{3}\left(\frac{T}{T_c}\right) t^{1/2}. \tag{6.9}$$

We note that the quantity in parentheses, T/T_c, approaches 1 at the critical temperature. As a result, it follows that the magnetization vanishes near T_c as

$$m \approx t^{1/2}.$$

In addition, the slope of m versus T goes to infinity as T_c is approached as $t^{-1/2}$.

These results verify that m does indeed have a power-law singularity near T_c. Recall that the magnetization critical exponent, β, is defined as

$$m \approx t^\beta.$$

Clearly then, the *mean-field* critical exponent for the magnetization is

$$\beta = \frac{1}{2}.$$

As mentioned previously, this result applies to all mean-field approximations.

Magnetization Critical Exponent: Exact 2-D

By way of contrast, let's take a look at the magnetization exponent from the exact result in 2-D. Referring to Equation 5.6, we can write the exact magnetization as follows:

$$m_{\text{exact}} = \left[1 - \frac{1}{\left(\sinh \left[\frac{2}{(T/T_c)T_c} \right] \right)^4} \right]^{1/8}.$$

Recall that the dimensionless critical temperature is $T_c = 1/K_c$, where, from Equation 5.3, we have

$$K_c = \frac{1}{2} \log(1 + \sqrt{2}) = 0.44068\ldots.$$

Noting that $T/T_c = 1 - t$, we can rewrite m_{exact} as follows:

$$m_{\text{exact}} = \left[1 - \frac{1}{\left(\sinh \left[\frac{2}{(1-t)T_c} \right] \right)^4} \right]^{1/8}.$$

Now, when $t = 0$ in this expression – which corresponds to the critical point – the sinh function is equal to 1. That is,

$$\sinh \left[\frac{2}{T_c} \right] = 1.$$

It follows that $m_{\text{exact}} = 0$ when $t = 0$, as expected.

Next, consider a straightforward Taylor series expansion of the sinh function for small values of t. This gives

$$\sinh\left[\frac{2}{(1-t)T_c}\right] = \sinh\left[\frac{2}{T_c}\right] + \left(\frac{2}{T_c}\right)\left(\cosh\left[\frac{2}{T_c}\right]\right)t + O(t^2)$$
$$= 1 + [\sqrt{2}\ln(1+\sqrt{2})]t + O(t^2)$$
$$= 1 + at + O(t^2).$$

Notice that we've simplified the expression by introducing the following constant:

$$a = \sqrt{2}\ln(1+\sqrt{2}). \tag{6.10}$$

This constant results from two identities. First,

$$\frac{2}{T_c} = \ln(1+\sqrt{2}).$$

Second, noting that $(\cosh[x])^2 - (\sinh[x])^2 = 1$, we have

$$\cosh\left[\frac{2}{T_c}\right] = \sqrt{1 + \left(\sinh\left[\frac{2}{T_c}\right]\right)^2} = \sqrt{2}.$$

These identities combine to give the expression for a in Equation 6.10.

Substituting the expansion of sinh into m_{exact}, we find

$$m_{exact} = \left[1 - \frac{1}{\left(1 + at + O(t^2)\right)^4}\right]^{1/8}$$
$$= [4at + O(t^2)]^{1/8} = (4a)^{1/8}t^{1/8} + \text{higher order.}$$

Clearly, the exact magnetization goes to zero as t raised to the 1/8 power, and hence for 2-D the exact magnetization exponent is

$$\beta = \frac{1}{8}.$$

This exponent applies to all ferromagnetic Ising models in 2-D, which includes the square lattice, triangular lattice, hexagonal lattice, and others.

Collected Critical Properties

We display the dimensionless critical temperature and magnetization exponent for various lattices in Table 6.2. The results listed as "Exact" come from explicit solutions for the 1-D and 2-D lattices, and from exact series expansion

Table 6.2 *Critical properties (dimensionless critical temperature* T_c *and magnetization exponent* β) *of various lattices from both exact and single-spin mean-field calculations. The "Exact" results for 3-D are the best estimates from exact series expansions, and are considered to be correct to this number of decimal places.*

Lattice	Dimension	T_c (Exact)	T_c (MF)	β (Exact)	β (MF)
Chain	1	0	2 —	—	½
Square	2	2.27	4 76% error	1/8	½
Triangular	2	3.64	6 65% error	1/8	½
Simple cubic	3	4.51	6 33% error	5/16	½
Face-centered cubic	3	9.79	12 23% error	5/16	½

analyses for the 3-D lattices. The mean-field results come from the single-spin approximation of this section.

Notice that the value of the mean-field magnetization exponent is the same for all dimensions. In contrast, the exact magnetization exponent varies with dimension, but it is the same for different lattices with the same dimension, even if those lattices have different types of interactions. This is the basis of the concept of "universality" in critical phenomena – critical exponents depend on the *global* symmetries of a system, not on the details of the *local* structure and interactions. On the other hand, there is no "universality" in the critical temperature – instead, it is sensitive not only to the type of lattice and the dimensionality, but also to the type of interactions on the lattice.

6.2 Mean-Field Energy and Specific Heat

Now that we've explored the magnetization of the Ising model in the mean-field approximation, and compared it with other results, we turn to the internal energy and the specific heat.

To begin, consider the spin, s_0, that forms the basis of our mean-field calculation. The energy, E_0, of this spin can be written as follows:

$$E_0 = -\frac{1}{2} q J m s_0.$$

Notice that we use a bond strength of $\frac{1}{2} J$ in this case, to avoid double counting the bonds. In contrast, when we wrote the Hamiltonian for this spin in the magnetization section, we used a bond strength of J, because we were including all the q interactions that influence s_0. In this case, we use $\frac{1}{2}J$ because we want to obtain the correct amount of energy per site of the lattice, and a lattice of N sites with periodic boundary conditions and coordination number q has $\frac{1}{2}q$ bonds per site.

Now, the average energy for the spin s_0, which we refer to as the internal energy per site, U/N, is

$$\frac{U}{N} = \langle E_0 \rangle = -\frac{1}{2}qJm\langle s_0 \rangle = -\frac{1}{2}qJm^2.$$

We've used the self-consistent relation, $<s_0> = m$, to obtain the final form of this expression. As a result, note that the internal energy is proportional to m^2, and hence it is zero for all temperatures above T_c. This means that the specific heat is also zero above the critical temperature. Both of these results are at odds with the actual behavior of the Ising model.

Below T_c the self-consistent solution with nonzero m gives a lower energy than the $m = 0$ solution. It follows that the nonzero-m solution is the one that describes the physical system below the critical temperature.

The next step is to take the temperature derivative of U/N to obtain the specific heat per site, C/N. Thus, we can write

$$\frac{C}{N} = \frac{d}{dT}\left(\frac{U}{N}\right) = -qJm\frac{dm}{dT}. \tag{6.11}$$

This expression is straightforward enough, but some care must be taken in evaluating the derivative of the magnetization.

First, we note that dm/dT can be written as follows:

$$\frac{dm}{dT} = \frac{d}{dT}\tanh\left[q\frac{J}{k_B T}m\right].$$

To carry out this derivative, we first take the derivative of the tanh function with respect to its argument, and then multiply that result by the derivative of the argument with respect to T. We will designate the argument with the label y, defined as follows:

$$y = q\frac{J}{k_B T}m.$$

Now, taking the derivative of the tanh function yields

$$\frac{d}{dy}\tanh[y] = \frac{(\cosh[y])^2 - (\sinh[y])^2}{(\cosh[y])^2} = 1 - (\tanh[y])^2.$$

It follows that dm/dT is given by

$$\frac{dm}{dT} = \frac{d}{dT}\tanh\left[q\frac{J}{k_B T}m\right] = \left(1 - \left(\tanh\left[q\frac{J}{k_B T}m\right]\right)^2\right)\frac{dy}{dT} = (1 - m^2)\frac{dy}{dT}.$$

In the last expression, we have once again used the self-consistent relation to replace the tanh function with m.

Next, we turn to the derivative of the argument of the tanh function. It can be evaluated as follows:

$$\frac{dy}{dT} = \frac{d}{dT}\left(q\frac{J}{k_B T}m\right) = \left[-q\frac{J}{k_B T^2}m + q\frac{J}{k_B T}\frac{dm}{dT}\right].$$

Combining all of these results yields

$$\frac{dm}{dT} = \frac{d}{dT}\tanh\left[q\frac{J}{k_B T}m\right] = (1 - m^2)\left[-q\frac{J}{k_B T^2}m + q\frac{J}{k_B T}\frac{dm}{dT}\right].$$

Notice that dm/dT appears on both sides of the equation. Rearranging, and using the expression $T_c = q\frac{J}{k_B}$, yields

$$\frac{dm}{dT} = -\frac{(1 - m^2)\left(\frac{T_c}{T}\right)\left(\frac{1}{T}\right)m}{1 - (1 - m^2)\left(\frac{T_c}{T}\right)} = -\frac{\left(\frac{T_c}{T}\right)\left(\frac{1}{T}\right)m}{\frac{1}{(1-m^2)} - \left(\frac{T_c}{T}\right)}.$$

Multiplying numerator and denominator by $(T/T_c)^2$ simplifies the expression further to the following:

$$\frac{dm}{dT} = -\frac{\left(\frac{1}{T_c}\right)m}{\frac{(T/T_c)^2}{(1-m^2)} - \left(\frac{T}{T_c}\right)}.$$

Finally, we use these results in our expression for C/N in Equation 6.11. The result is

$$\frac{C}{N} = -qJm\frac{dm}{dT} = \frac{\left(\frac{qJ}{T_c}\right)m^2}{\frac{(T/T_c)^2}{(1-m^2)} - \left(\frac{T}{T_c}\right)} = \frac{k_B m^2}{\frac{(T/T_c)^2}{(1-m^2)} - \left(\frac{T}{T_c}\right)}.$$

Converting this to a dimensionless form, making it convenient for plotting, and relabeling with the convenient shorthand name c, we have

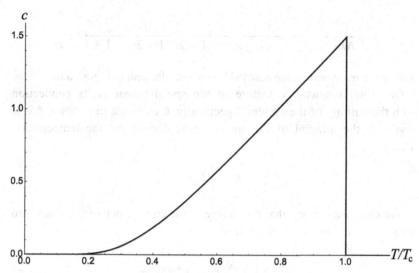

Figure 6.6 The dimensionless specific heat per site, $c = C/Nk_B$, for the Ising model in the single-spin mean-field approximation. It reaches a maximum value of 1.5 at $T = T_c$ and is zero for temperatures above T_c.

$$c = \frac{C}{Nk_B} = \frac{m^2}{\frac{(T/T_c)^2}{(1-m^2)} - \left(\frac{T}{T_c}\right)}. \tag{6.12}$$

This result is shown in Figure 6.6.

Notice that the specific heat reaches a maximum value at T_c and then is identically equal to zero at higher temperatures. In contrast, the exact specific heat for the 2-D square lattice diverges to infinity at the critical temperature and is nonzero for all other temperatures. Thus, some of the general features of the exact result are captured by the mean-field approximation – namely, a peak in the specific heat at T_c – though others are decidedly incorrect.

The value of the peak of the specific heat can be obtained by referring to the asymptotic form of the magnetization near the critical point, as presented in Equation 6.9. Specifically, we can see that as the reduced temperature, $t = 1 - T/T_c$, goes to zero, $t \to 0$, the square of the magnetization approaches the following:

$$m^2 \to 3t.$$

Substituting this result and the expression for t into Equation 6.12 for the reduced specific heat, we find

$$c = \frac{C}{Nk_B} \rightarrow \frac{3t}{\frac{(1-t)^2}{1-3t} - (1-t)} \rightarrow \frac{3t}{(1-2t)(1+3t) - 1 + t} \rightarrow \frac{3}{2}.$$

This is the maximum value reached by the specific heat in Figure 6.6.

One final noteworthy feature of the specific heat is its connection with the entropy of the system. Specifically, the change in entropy, ΔS, is given by the integral of the specific heat divided by the temperature. That is,

$$\Delta S = \int \frac{C}{T} dT.$$

In our case, we can say that the change in the entropy per site, s, from zero temperature to infinite temperature is

$$\Delta s = \int_0^\infty \frac{C/N}{T} dT = \int_0^{T_c} \frac{C/N}{T} dT. \tag{6.13}$$

The upper limit in the second integral has been replaced with T_c, since the specific heat is zero for all higher temperatures. Using the expression for C/N given in Equation 6.11, and simplifying the integral, we find

$$\Delta s = -\int_0^{T_c} qJ \frac{m}{T} \frac{dm}{dT} dT = -\int_1^0 qJ \frac{m}{T} dm = \int_0^1 qJ \frac{m}{T} dm.$$

Notice that the integral over m runs from 0, corresponding to the disordered state, to 1, corresponding to the ground state. Finally, by inverting our self-consistent relation, we obtain

$$qJ \frac{m}{T} = k_B \tanh^{-1}[m].$$

Substituting this into the integral for Δs yields

$$\Delta s = k_B \int_0^1 \tanh^{-1}[m] dm = k_B \ln 2. \tag{6.14}$$

To be clear, note that the integral is over the inverse hyperbolic tangent, $\tanh^{-1}$, *not* the hyperbolic tangent to the -1 power.

We see that the change in the entropy per site for the Ising model from the ground state to the disordered state is simply k_B times the natural log of 2. The physical significance of this result becomes apparent when we consider the entropy of the two limiting cases. In the ground state, at $T = 0$, there are just two states – all spins up or all spins down. Thus, the entropy per site in this case is

$$s = \frac{1}{N} k_B \ln 2 \underset{N \to \infty}{\to} 0.$$

As expected, the entropy per site is zero in the ground state. For the totally disordered state, at infinite temperature, each spin is equally likely to be in either of its two possible states. It follows that there are 2^N states in this case, and hence the entropy per site is

$$s = \frac{1}{N} k_B \ln 2^N = k_B \ln 2.$$

As we saw earlier, in Equation 6.13, the change in entropy from $T = 0$ to $T = \infty$ is the same as the change in entropy from $T = 0$ to $T = T_c$, since the specific heat is zero for all temperatures higher than T_c. Thus, the result in Equation 6.14 agrees with our expectations for the Ising model.

6.3 The Landau–Ginzburg Approach to Mean-Field Theory

In this section, we investigate a *phenomenological* way of looking at critical behavior, known to as the Landau–Ginzburg method. By phenomenological, we mean that this approach isn't based on a quantitative, first-principles calculation of a specific system, as was the case with the single-spin mean-field theory. Instead, the Landau–Ginzburg method is based on the *qualitative properties* one would expect for a system near a critical point. As such, it can yield valuable insight without a detailed calculation.

Zero-Field Properties

To begin, we note that near a critical point, the order parameter – which is m in the case of an Ising magnet – is arbitrarily small. Hence, it seems reasonable to consider an expansion of the free energy in powers of m. We know that this may be problematic, owing to singularities in the system, but it makes sense to start here.

In the case of zero magnetic field, where the system is unchanged by reversing all the spins, we can expand the reduced free energy per site, f, as follows:

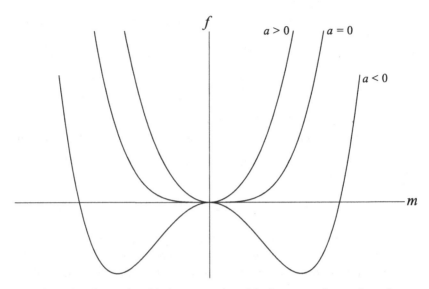

Figure 6.7 The Landau–Ginzburg expansion of the free energy for $a > 0$, $a = 0$, and $a < 0$. In all cases, $b > 0$. For a < 0, the finite-m solutions minimize the free energy.

$$f = am^2 + bm^4.$$

Note that we include only even powers of m in our expansion – this ensures that f is invariant under a change in sign of m. In addition, we truncate the expansion at the fourth order since this is high enough to yield the critical behavior of interest. We will have more to say about the multiplicative terms a and b as we continue our discussion.

Now, the basic idea in this approach is that a system in thermodynamic equilibrium seeks to minimize its free energy. To look for the minimum values of the free energy, we plot f in Figure 6.7 for three different cases. In each case, we assume that b is greater than zero – this bounds the minima to small values of m and keeps the free energy from going to minus infinity for large m. The value of a can change sign, however. For $a > 0$ and $a = 0$ the free energy has a single minimum, which is at $m = 0$; for $a < 0$ the free energy has two symmetrically placed minima at positive and negative values of m. The value of the free energy at the nonzero-m solutions is always less than it is at $m = 0$, which is actually a local maximum for $a < 0$.

We've seen this kind of behavior before. In fact, when we studied the self-consistent condition for the single-spin mean-field calculation, Equation 6.5, we saw that $m = 0$ is always a solution. We also saw that for temperatures below the critical temperature there are two new solutions, with symmetrically placed positive and negative values of m. At the time, we mentioned that these finite-m solutions minimize the free energy, and we can see now that this is indeed the case in Figure 6.7.

In addition, it follows that we can identify the critical temperature in the Landau–Ginzburg approach with the value $a = 0$. After all, that is where the finite-m solutions first begin to appear. The simplest assumption for the dependence of a on temperature near the critical point, $t = 0$, is that it varies linearly. Hence, we can write

$$a = -a't.$$

In this expression, $a' > 0$ is the negative of the rate of change of a near $t = 0$.

To explore the behavior of the magnetization m near the critical point, and to determine the magnetization critical exponent, we look for minima of the free energy function f. To do this, we take the derivative of f with respect to m and set the result equal to zero. This yields

$$\frac{df}{dm} = 2am + 4bm^3 = 0.$$

Clearly, $m = 0$ is always a solution, as expected, and it is the physical solution for $a \geq 0$; that is, for temperatures above the critical temperature. In addition, the finite-m solutions, which apply for temperatures below the critical temperature, are given by the following:

$$m = \sqrt{-\frac{2a}{4b}} = \sqrt{\frac{2a't}{4b}}.$$

It follows that m vanishes like the square root of t:

$$m \sim t^{1/2}.$$

As a result, the critical exponent for the Landau–Ginzburg approach is

$$\beta = \frac{1}{2}.$$

This is exactly the same magnetization critical exponent as in other mean-field approaches. In general, all versions of mean-field theory predict all the same exponents.

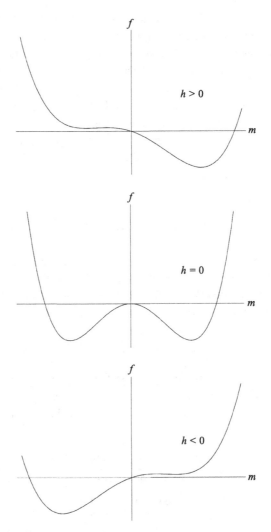

Figure 6.8 The Landau–Ginzburg expansion of the free energy with a magnetic field term, $-hm$. We present the following three cases: (a) $h > 0$; (b) $h = 0$; (c) $h < 0$.

Finite-Field Properties

Let's take a look now at the Landau–Ginzburg free energy in the case that the system has a finite magnetic field, h. This adds the following term to the expression for f:

$$-hm.$$

This term breaks the symmetry about $m = 0$, as one would expect, and follows from taking the hs_i term in the Hamiltonian (Equation 4.37) and replacing s_i with its mean value, m.

In Figure 6.8 we show plots of the free energy f for the following cases: (a) $h > 0$; (b) $h = 0$; and (c) $h < 0$. When $h > 0$ there is only one lowest minimum in the free energy, and it occurs at a positive value of the magnetization, $m > 0$. It follows that the system responds to the finite magnetic field with a finite magnetization.

When the magnetic field is zero, $h = 0$, the free energy is again symmetric about $m = 0$. For temperatures below the critical temperature there are two finite-m minima, and they have the same value of the free energy. Thus, we expect the system to have a *coexistence* of a positive-m state and a negative-m state in zero field.

Finally, with $h < 0$ the free energy again has only a single lowest minimum, which this time occurs at $m < 0$. As h is lowered from positive values to zero, and then to negative values, the magnetization jumps discontinuously – a first-order, or discontinuous, phase transition – from the positive-m state to the negative-m state. There is no critical point when the magnetic field is finite.

Thus, the Landau–Ginzburg approach provides a useful way of visualizing the behavior of a system as it undergoes various phase transitions. It also gives predictions for the critical exponents, and these are in agreement with other mean-field theories.

6.4 Problems

6.1 Consider $T/T_c = 0.5$ in the single-spin mean-field calculation. Starting with $m_0 = 0.25$, find (a) m_1 and (b) m_2.

6.2 Find m^* for $T/T_c = 0.75$ in the single-spin mean-field calculation.

6.3 Find (a) the magnetization, m, and (b) the reduced specific heat per site, c, for $T/T_c = 0.85$ in the single-spin mean-field calculation.

7

Position-Space Renormalization-Group Techniques

Before the early 1970s there were only a few ways to study the Ising model. One was to do an exact solution (Chapters 4 and 5), in those few cases where it is possible to do so. Other approaches involved doing an exact series expansion about high or low temperatures (Chapter 5), or carrying out an approximate mean-field calculation (Chapter 6). All of this changed in 1971, however, when Kenneth Wilson introduced the renormalization-group (RG) approach. The implications of this breakthrough were immediately recognized by researchers in the field, and Wilson and the RG technique were awarded the Nobel Prize in physics soon thereafter.

One of the distinguishing features of RG methods is that they explicitly include the effects of fluctuations, as opposed to mean-field calculations that focus solely on average values. In addition, the RG approach gives a natural understanding of the universality that is seen in critical phenomena in general, and in critical exponents in particular. In many respects, the RG approach gives a deeper understanding not only of the Ising model itself, but of all aspects of critical phenomena.

The original version of the renormalization-group method was implemented in momentum space – which is a bit like studying a system with Fourier transforms – and is beyond the scope of this presentation. Following that, various investigators, including the author, his coworkers, and many others, extended the approach to position space, which is more intuitive in many ways, and is certainly much easier to visualize. As we shall see, position-space renormalization-group (PSRG) calculations can be done exactly in some cases, while in others we will need to construct an approximate RG transformation. Either way, we will find results that take us well beyond mean-field treatments.

We present the basics of position-space renormalization group methods in this chapter. We will also explain the origin of the terms "renormalization" and "group" in the RG part of the name. Finally, it's worth noting that the calculations used in PSRG are similar to those presented in Chapters 2 and 3, especially those used in the partial-summation transformations. It may be useful to briefly review those chapters before continuing with the present discussion.

7.1 Exact Position-Space Renormalization-Group Calculations in One Dimension

We start with our old friend, the 1-D Ising chain. We've solved this system in several different ways, each illustrating a different kind of approach. In this chapter, we use the 1-D Ising model as our first example of a PSRG calculation.

To start, consider a 1-D Ising chain of N sites, as shown in Figure 7.1 (a). We are interested in an infinite system, and so we will eventually let $N \to \infty$. Nearest neighbors interact with a dimensionless coupling K, and hence the reduced Hamiltonian for the system is

$$-\beta H = K \sum_{\langle i,j \rangle} s_i s_j.$$

As usual, the summation is over all nearest neighbors on the lattice.

The basis for this exact PSRG calculation is referred to as a *decimation transformation*. This simply means that we sum over the configurations of a *fraction* of the spins on the original lattice. In this case we sum over half of the original spins, as indicated by the Xs in Figure 7.1 (a) – this is referred to as

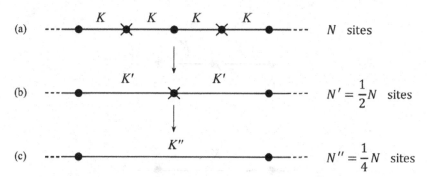

Figure 7.1 Decimation transformation for a 1-D PSRG calculation. Half of the spins are summed over in each step of the transformation, generating a new coupling and resetting the zero level of the system.

decimating the lattice. These summations produce a new (primed) lattice with half the number of sites, $N' = N/2$, and with an effective (or *renormalized*) coupling K', as illustrated in Figure 7.1 (b). The procedure is then repeated on the primed lattice, producing a double-primed lattice with $N'' = N/4$ sites and a new renormalized coupling K''. As we shall see, this procedure can be repeated an infinite number of times to produce a solution to the original infinite chain.

The basic unit of the transformation is shown in Figure 7.2. Notice that it is virtually the same as the partial-summation transformation we considered in Section 3.3. As was done there, we sum over the configurations of s_2 to generate an effective coupling between s_1 and s_3. This also generates a zero-spin term, K_0', which resets the zero level of the system (*renormalizes* it) and will be of considerable importance in calculating the free energy, as we shall see.

To carry out the transformation, we note that the reduced Hamiltonian of the original system in Figure 7.2 is

$$-\beta H = K(s_1 s_2 + s_2 s_3).$$

Similarly, the reduced Hamiltonian of the primed, or renormalized, system is

$$-\beta H' = K' s_1 s_3 + K_0'.$$

Notice that we have only K_0' in this Hamiltonian, whereas we had $2K_0'$ in Section 3.3. The difference is that in the former case we had one site on either end of the basic unit, and hence one K_0' for each of those sites. In this case, the sites on either end are shared equally with their neighbors, thus we have two sites contributing $K_0'/2$ for a total contribution of K_0'.

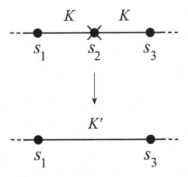

Figure 7.2 The basic unit of the decimation transformation from Figure 7.1.

Preserving the Partition Function

Now, the basic idea of this transformation is to preserve the partition function as we go from the original lattice to the primed lattice. Specifically, we want the partition function for the spins s_1 and s_3 with the renormalized Hamiltonian H' to be equal to the partition function of s_1, s_2, and s_3 with the original Hamiltonian H. Stated mathematically, this requirement is as follows:

$$\sum_{\{s_1,s_3\}} e^{-\beta H'} = \sum_{\{s_1,s_3\}} \sum_{\{s_2\}} e^{-\beta H}.$$

Equivalently, we can say that the following equality must hold for any given configuration of s_1 and s_3:

$$e^{-\beta H'} = \sum_{\{s_2\}} e^{-\beta H}.$$

For example, suppose we let $s_1 = +1$, $s_3 = +1$. This yields the condition

$$e^{K'+K_0'} = \sum_{\{s_2\}} e^{K(s_2+s_2)} = e^{2K} + e^{-2K} = Z_{++}.$$

In this expression we've introduced the notation Z_{++} to indicate the *partial partition function* for the case $s_1 = +1$, $s_3 = +1$. Taking the log of both sides yields

$$K' + K_0' = \ln[Z_{++}] = \ln\left[e^{2K} + e^{-2K}\right].$$

This equation gives one relation for the two unknowns, K' and K_0'.

Next, consider the case $s_1 = +1$, $s_3 = -1$. Carrying out a similar calculation, we find

$$e^{-K'+K_0'} = \sum_{\{s_2\}} e^{K(s_2-s_2)} = 2 = Z_{+-}.$$

Again, taking the log of both sides yields

$$-K' + K_0' = \ln[Z_{+-}] = \ln[2].$$

This is a second equation for K' and K_0'. Note that nothing new is added by considering the cases $s_1 = -1$, $s_3 = +1$ and $s_1 = -1$, $s_3 = -1$, since the system is symmetric under spin reversal.

It follows that we now have two independent equations for the two unknowns:

$$K' + K'_0 = \ln[Z_{++}] = \ln[e^{2K} + e^{-2K}]$$
$$-K' + K'_0 = \ln[Z_{+-}] = \ln[2].$$

These equations can be solved to give

$$K' = \frac{1}{2}\ln\left[\frac{Z_{++}}{Z_{+-}}\right] = \frac{1}{2}\ln[\cosh(2K)]$$

$$K'_0 = \frac{1}{2}\ln[Z_{++}Z_{+-}] = \frac{1}{2}\ln[4\cosh(2K)] = \ln 2 + K'. \tag{7.1}$$

With this renormalized coupling, K', and renormalized zero level of the Hamiltonian, K'_0, the partition function of the primed system is the same as the partition function of the original system, as desired.

Renormalized Coupling

The renormalized coupling $K'(K)$ is displayed in Figure 7.3. Notice that $K'(K)$ is less than K for all finite values of K. This makes sense physically. After all, if the spins s_1 and s_2 are correlated to a certain extent due to the effect of the coupling K, then the spins s_1 and s_3 – which are less correlated because of their

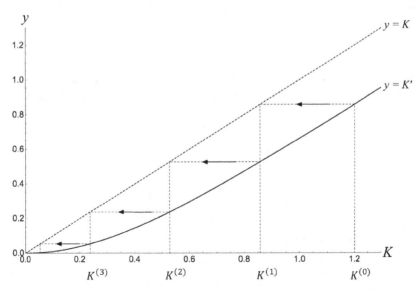

Figure 7.3 The renormalized coupling K' as a function of K. Notice that K' is always less than K. The cobwebbing between $y = K$ and $y = K'$ shows that the flow of the coupling is toward $K = 0$, as indicated with the arrows.

Table 7.1 *Successive iterations of the transformation for* K', *given by the top expression in Equation 7.1. Notice the rapid convergence toward* K = 0.

n	$K(n)$
0	1.2
1	0.8575
2	0.5269
3	0.2377
4	0.0545
5	0.0030
6	0.0000
.	.
.	.
.	.
∞	0

greater separation – will have an effective coupling K' that is less than K. The only exception is when there is no coupling at all, $K = 0$, in which case the renormalized coupling will also be zero. Thus, $K = 0$ is a fixed point of the transformation; that is, $K'(0) = 0$.

Also shown in Figure 7.3 is the way that successive iterations of the transformation can be visualized in terms of cobwebbing. To assist with this, we designate the initial value of K with the notation $K = K^{(0)}$. The first iteration is $K' = K^{(1)}$, the second iteration is $K'' = K^{(2)}$, and the nth iteration is $K^{(n)}$. For example, consider the initial coupling $K^{(0)} = 1.2$. To cobweb with this value in Figure 7.3, we start on the K axis at $K = K^{(0)}$, then go vertically upward to the $y = K'$ curve. We then go horizontally to the $y = K$ line, and then back down to $y = K'$ at the value $K' = K^{(1)} = 0.8575$. This is the first step in the cobweb. Continuing this process takes us closer and closer to the *attractive fixed point* at $K = 0$. We can visualize this as an RG "*flow*" of the coupling K toward the fixed point, as indicated by the arrows on the cobwebbing. The full sequence of values of $K^{(n)}$ for this starting point is given in Table 7.1.

Free Energy

We turn now to K_0' and the connection it has to the free energy of the system. To begin, recall that the decimation transformation preserves the partition function Z as we go from one lattice to the next. For the original lattice, with $N \to \infty$ sites and coupling K, the partition function is

$$Z_N(K) = \sum_{\{s_1, s_2, s_3, \ldots\}} e^{K(s_1 s_2 + s_2 s_3 + \ldots)}. \qquad (7.2)$$

Notice that the summation is over the configurations of all N spins.

The partition function for the primed lattice, which has been set equal to the original partition function, $Z_N(K)$, is basically the same as the result in Equation 7.2, although now for a lattice with $N' = N/2 \to \infty$ sites and a new coupling K'. It also has a K'_0 term for each site of the primed lattice; this term resets the zero level, as discussed previously. Thus, we can write $Z_N(K)$ as follows:

$$Z_N(K) = \sum_{\{s_1, s_3, s_5, \ldots\}} e^{K'(s_1 s_3 + s_3 s_5 + \ldots) + N'K'_0}.$$

In this case, the summation is over the *infinite* number of *odd* spins on the lattice.

Next, let's pull the zero-level term out in front of the expression for the partition function:

$$Z_N(K) = e^{N'K'_0} \sum_{\{s_1, s_3, s_5, \ldots\}} e^{K'(s_1 s_3 + s_3 s_5 + \ldots)}.$$

Though the numbering of the sites and the coupling are different, the remaining summation is precisely the same as in Equation 7.2 – an infinite sum over N' spins with nearest-neighbor interactions K'. Thus, we can write

$$Z_N(K) = e^{N'K'_0} Z_{N'}(K'). \qquad (7.3)$$

This is the detailed connection between the two lattices.

To convert the preceding relationship to an expression for the free energies, recall that the reduced free energy per site, f, is equal to the log of the partition function divided by the number of sites. Thus, for the original lattice we have

$$f(K) = \frac{1}{N} \ln Z_N(K).$$

Using Equation 7.3, we can write $f(K)$ as follows:

$$f(K) = \frac{1}{N} N'K'_0 + \frac{1}{N} \ln Z_{N'}(K') = \frac{N'}{N} K'_0 + \frac{N'}{N} \frac{1}{N'} \ln Z_{N'}(K').$$

Noting that $N' = N/2$, we have

$$f(K) = \frac{1}{2} K'_0 + \frac{1}{2} \frac{1}{N'} \ln Z_{N'}(K').$$

Finally, the last term in the previous equation is simply one half the reduced free energy per site of the primed lattice. Therefore,

$$f(K) = \frac{1}{2}K_0' + \frac{1}{2}f(K'). \tag{7.4}$$

The two free energies in Equation 7.4, $f(K)$ and $f(K')$, have the same functional form, since they both represent an infinite lattice with nearest-neighbor interactions only; they simply have different arguments.

To check this expression, we start with the simplest possible test – namely, to verify that it is valid for $K = 0$. In this case, $K' = 0$, $K_0' = \ln(2)$, and $f(K) = \ln(2)$. Combining these results in Equation 7.4 yields

$$\ln(2) = \frac{1}{2}\ln(2) + \frac{1}{2}\ln(2).$$

This shows that we have indeed included the correct counting factors in our transformation. The expression in Equation 7.4 can also be verified for finite values of K and K', which we explore in the end-of-chapter problems.

Free Energy: Numerical Summation

Equation 7.4 gives a rigorous connection between the free energy of the original lattice and the free energy of the corresponding primed lattice. It can also be used as the basis for a direct numerical calculation of the free energy. To see how, let's extend Equation 7.4 one step further. This yields the following:

$$f(K) = \frac{1}{2}K_0' + \frac{1}{2}f(K') = \frac{1}{2}K_0'(K) + \frac{1}{4}K_0'(K') + \frac{1}{4}f(K'').$$

Continuing this process gives

$$f(K) = \frac{1}{2}K_0'(K) + \frac{1}{4}K_0'(K') + \frac{1}{8}K_0'(K'') + \dots.$$

Using the notation introduced earlier, with numbered superscripts, we can write the calculation of the free energy as an infinite sum:

$$f(K) = \frac{1}{2}K_0'\left(K^{(0)}\right) + \frac{1}{4}K_0'\left(K^{(1)}\right) + \frac{1}{8}K_0'\left(K^{(2)}\right) + \dots$$

$$= \sum_{i=0}^{\infty}\left(\frac{1}{2^{i+1}}\right)K_0'\left(K^{(i)}\right). \tag{7.5}$$

Carrying out this summation, which actually converges rather quickly, is a convenient and useful way to calculate the free energy.

Table 7.2 *Successive iterations showing the*
convergence of the free energy to its exact value.

n	$K^{(n)}$	$\sum_{i=0}^{n} \left(\frac{1}{2^{i+1}}\right) K_0' \left(K^{(i)}\right)$
0	1.2	0.7753
1	0.8575	1.0803
2	0.5269	1.1967
3	0.2377	1.2434
4	0.0545	1.2652
5	0.0030	1.2760
6	0.0000	1.2814
.	.	.
.	.	.
.	.	.
∞	0	1.2868

As an example, consider starting with the coupling $K = K^{(0)} = 1.2$. The first contribution to the free energy, from the first renormalization-group step, is

$$\frac{1}{2} K_0' \left(K^{(0)}\right) = 0.7753.$$

This first step also produces the renormalized coupling we obtained earlier (see Table 7.1) when discussing cobwebbing:

$$K^{(1)} = 0.8575.$$

Using this value for $K^{(1)}$, and noting that $K_0'(K^{(1)}) = 1.2200$, we can calculate the sum of the first and second contributions to the free energy:

$$\frac{1}{2} K_0' \left(K^{(0)}\right) + \frac{1}{4} K_0' \left(K^{(1)}\right) = 1.0803.$$

Continuing this process gives closer and closer approximations to the exact free energy, $f_{\text{exact}}(1.2) = \ln(2) + \ln[\cosh(1.2)] = 1.2868$. The results of the first several steps are presented in Table 7.2. Notice the rapid convergence of the coupling to the fixed point at $K = 0$, and the correspondingly rapid convergence to the exact free energy.

Free Energy: Exact Summation

The numerical calculation of the free energy presented in Table 7.2 can be extended analytically to give the exact free energy as a general function of K.

The first step in this process is to write K_0' in terms of $K^{(n)}$. Referring to the bottom expression in Equation 7.1, we have

$$K_0'(K) = \ln 2 + K'.$$

In terms of superscripts, this can be written as follows:

$$K_0'\left(K^{(n)}\right) = \ln 2 + K^{(n+1)}.$$

Using this in the middle part of the expression for the free energy in Equation 7.5 yields

$$f(K) = \frac{1}{2}\left(\ln 2 + K^{(1)}\right) + \frac{1}{4}\left(\ln 2 + K^{(2)}\right) + \frac{1}{8}\left(\ln 2 + K^{(3)}\right) + \cdots$$

$$= \left(\frac{1}{2} + \frac{1}{4} + \frac{1}{8} + \cdots\right)\ln(2) + \frac{1}{2}K^{(1)} + \frac{1}{4}K^{(2)} + \frac{1}{8}K^{(3)} + \cdots.$$

The first summation, the one that multiplies ln (2), is a geometric series that sums to 1. The remaining summation, when combined with ln (2) and expressed in terms of superscripts, gives

$$f(K) = \ln(2) + \sum_{n=1}^{\infty} \frac{1}{2^n} K^{(n)}. \tag{7.6}$$

This is the free energy given in terms of K and its successive iterations.

The next step is to reexpress K' in terms of tanh of K. This may seem like an odd detour at this point, but the result is well worth the effort. We let v represent the tanh. That is,

$$v = \tanh(K) = \frac{e^K - e^{-K}}{e^K + e^{-K}}.$$

Similarly,

$$v' = \tanh(K') = \frac{e^{K'} - e^{-K'}}{e^{K'} + e^{-K'}}.$$

Referring to the top expression in Equation 7.1 allows us to rewrite v' in terms of K:

$$v' = \frac{\left(\cosh(2K)\right)^{1/2} - \left(\cosh(2K)\right)^{-1/2}}{\left(\cosh(2K)\right)^{1/2} + \left(\cosh(2K)\right)^{-1/2}} = \frac{\cosh(2K) - 1}{\cosh(2K) + 1}.$$

Expanding this result in terms of exponentials yields

$$v' = \frac{e^{2K} + e^{-2K} - 2}{e^{2K} + e^{-2K} + 2} = \frac{(e^K - e^{-K})^2}{(e^K + e^{-K})^2} = \left(\tanh(K)\right)^2 = v^2.$$

Thus, renormalizing v simply yields v^2. It follows that $v'' = v'^2 = v^4$. Extending these results, we have

$$v^{(n)} = v^{2^n}.$$

This makes it quite easy to write v for any given iteration.

Inverting the expression for v, to solve once again for K, we have

$$K = K^{(0)} = \frac{1}{2} \ln \left[\frac{1 + v}{1 - v} \right].$$

Generalizing to the nth iteration

$$K^{(n)} = \frac{1}{2} \ln \left[\frac{1 + v^{2^n}}{1 - v^{2^n}} \right].$$

Now, we can substitute this back into our expression for the free energy in Equation 7.6:

$$f(K) = \ln(2) + \sum_{n=1}^{\infty} \frac{1}{2^n} \left(\frac{1}{2} \ln \left[\frac{1 + v^{2^n}}{1 - v^{2^n}} \right] \right).$$

The sum of logs is more easily handled as the log of a product. Thus,

$$f(K) = \ln(2) + \ln \left[\prod_{n=1}^{\infty} \left(\frac{1 + v^{2^n}}{1 - v^{2^n}} \right)^{\frac{1}{2^{n+1}}} \right].$$

This is a new and quite unexpected form for the free energy. But we're not done just yet – we still have to evaluate the infinite product.

Actually, this is fairly easy to do when all is said and done. First, we expand the infinite product as follows:

$$\prod_{n=1}^{\infty} \left(\frac{1 + v^{2^n}}{1 - v^{2^n}} \right)^{\frac{1}{2^{n+1}}} = \left(\frac{1 + v^2}{1 - v^2} \right)^{\frac{1}{4}} \left(\frac{1 + v^4}{1 - v^4} \right)^{\frac{1}{8}} \left(\frac{1 + v^8}{1 - v^8} \right)^{\frac{1}{16}} \cdots$$

Next, we gather together powers of $(1 + v^2)$, $(1 - v^2)$, $(1 + v^4)$, and so on. This gives

$$\prod_{n=1}^{\infty} \left(\frac{1+v^{2^n}}{1-v^{2^n}}\right)^{\frac{1}{2^{n+1}}} = \left(\frac{1+v^2}{1-v^2}\right)^{\frac{1}{4}} \left(\frac{1+v^4}{(1+v^2)(1-v^2)}\right)^{\frac{1}{8}} \left(\frac{1+v^8}{(1+v^4)(1+v^2)(1-v^2)}\right)^{\frac{1}{16}} \cdots$$

$$= (1+v^2)^{\frac{1}{4}-\frac{1}{8}-\frac{1}{16}-\cdots}(1-v^2)^{-\frac{1}{4}-\frac{1}{8}-\frac{1}{16}-\cdots}(1+v^4)^{\frac{1}{8}-\frac{1}{16}-\cdots}\cdots$$

$$= (1+v^2)^0(1-v^2)^{-\frac{1}{2}}(1+v^4)^0\cdots = \frac{1}{\sqrt{1-v^2}}.$$

So, the entire infinite product collapses to just a single term in the end.

Returning to our expression for the free energy, we can now write it in the following form:

$$f(K) = \ln(2) + \ln\left[\prod_{n=1}^{\infty}\left(\frac{1+v^{2^n}}{1-v^{2^n}}\right)^{\frac{1}{2^{n+1}}}\right] = \ln(2) + \ln\left[\frac{1}{\sqrt{1-v^2}}\right]$$

$$= \ln(2) + \ln[\cosh(K)].$$

This is the exact result given previously in Equation 4.4. It's been a long way around to arrive at this result, but it's gratifying to verify that the PSRG transformation is indeed exact, and that it produces a complete solution of the system.

Finite Magnetic Field

All of our results so far have been for the case of zero magnetic field, $h = 0$. Adding a magnetic field is quite straightforward, however, and we shall do so now.

The basic unit for the decimation transformation with finite h takes the form shown in Figure 7.4. Notice that we add h to the central spin s_2, but only $\frac{1}{2}h$ to the outer spins s_1 and s_3. This is because these spins are shared equally with

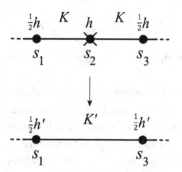

Figure 7.4 The basic unit for decimation with finite magnetic field, h.

their neighbors to the left and right. The same holds true for the primed lattice and the primed magnetic field, h'.

It follows that the reduced Hamiltonians for these sections of the original and primed lattices are as follows:

$$-\beta H = K(s_1 s_2 + s_2 s_3) + h\left(\frac{1}{2}s_1 + s_2 + \frac{1}{2}s_3\right)$$

$$-\beta H' = K' s_1 s_3 + \frac{1}{2}h'(s_1 + s_3) + K_0'.$$

Similarly, applying the procedure described earlier in this chapter, we can determine expressions for the renormalized terms in $-\beta H'$. We find the following:

$$K' + h' + K_0' = \ln[Z_{++}] = \ln[e^{2K+2h} + e^{-2K}]$$

$$-K' + 0 + K_0' = \ln[Z_{+-}] = \ln[e^{h} + e^{-h}]$$

$$-K' + 0 + K_0' = \ln[Z_{-+}] = \ln[e^{h} + e^{-h}]$$

$$K' - h' + K_0' = \ln[Z_{--}] = \ln[e^{2K-2h} + e^{-2K}].$$

There are really only three independent equations here, since the middle two are equal due to the symmetry of the system. These three equations determine the three renormalized terms, K', h', and K_0'. In fact, we can rearrange and solve for these terms as follows:

$$K' = \frac{1}{4}\ln\left[\frac{Z_{++}Z_{--}}{Z_{+-}Z_{-+}}\right]$$

$$h' = \frac{1}{2}\ln\left[\frac{Z_{++}}{Z_{--}}\right]$$

$$K_0' = \frac{1}{4}\ln[Z_{++}Z_{+-}Z_{-+}Z_{--}].$$

These equations fully determine the decimation transformation for general values of h.

Notice that this transformation takes the original values of K and h and generates the primed values K' and h'. The K_0' terms are basically just coming along for the ride, and will be used shortly to calculate the free energy. Thus, we can think of the transformation as a "*group operation*," in the sense that it takes the two values (K, h) and transforms – or renormalizes – them to the two new values (K', h'). It is from these considerations that the name renormalization *group* (RG) originated.

As before, the free energy of the system can be calculated with the K_0' terms in the transformation. Specifically, the free energy transforms as follows:

$$f(K, h) = \frac{1}{2} K_0'(K, h) + \frac{1}{2} f(K', h').$$

Expanding on this process yields

$$f(K, h) = \frac{1}{2} K_0'(K, h) + \frac{1}{4} K_0'(K', h') + \frac{1}{8} K_0'(K'', h'') + \ldots.$$

This can be used to give a numerical solution for the free energy.

As an example, let's start with the values $K = 1.2$ and $h = 1.2$. Carrying out the PSRG procedure, often referred to as "turning the crank," we find the values given in Table 7.3. Notice that K rapidly flows to $K = 0$, at which point h stops changing. Thus, any point on the line $K = 0$ is a fixed point, regardless of the value of h. In addition, notice that the free energy converges to the value 2.4008, precisely the same as the value given by the exact solution:

$$f_{\text{exact}}(K, h) = \ln \left[e^K \cosh(h) + \sqrt{e^{-2K} + e^{2K} [\sinh(h)]^2} \right].$$

This can be repeated with any values of K and h. Flows of this transformation are shown in Figure 7.5.

Notice that the flows go toward larger values of h and smaller values of K until $K = 0$ is reached, at which point there is no further change in K and h. Even so, successive iterations continue to add more increments to the free

Table 7.3 *Renormalization-group flow for the starting values* $K = 1.2$ *and* $h = 1.2$. *Also shown is the convergence of the free energy to its exact value.*

n	$K^{(n)}$	$h^{(n)}$	$\sum_{i=0}^{n} \left(\frac{1}{2^{i+1}} \right) K_0^{(i+1)}$
0	1.2	1.2	0.9327
1	0.5785	2.3570	1.5285
2	0.0175	3.4709	1.9625
3	0	3.5058	2.1817
4	0	3.5058	2.2912
5	0	3.5058	2.3460
.	.	.	.
.	.	.	.
∞	0	3.5058	2.4008

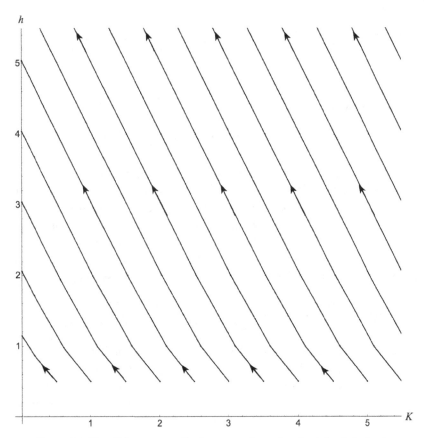

Figure 7.5 Visualizing renormalization-group flows for the 1-D chain in the K-h plane.

energy, in terms of K_0', giving convergence to any desired number of significant digits.

7.2 The Migdal–Kadanoff Renormalization-Group Transformation

The PSRG methods presented in the previous section can be applied to two- and higher-dimensional systems with a transformation introduced by A. A. Migdal and L. P. Kadanoff. The Migdal–Kadanoff (MK) transformation, as it is known, consists of two parts. The first part is a "restructuring" of the original lattice, in which bonds are moved to new locations. This step is

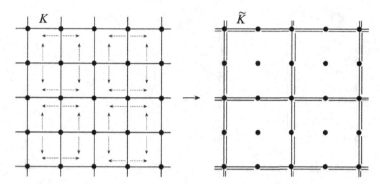

Figure 7.6 The first step of the MK transformation. This is the restructuring, or bond-moving step, in which the bonds inside the 2 × 2 squares on the original lattice (left) are moved outwards to the edges of the squares on the new lattice (right). The original bond strength is K, the number of sites is N on both lattices, and the restructured bond strength is $\widetilde{K} = 2K$.

approximate in nature. The second part is an exact decimation transformation, just like the ones we've already studied. Combined together, these steps produce an approximate transformation that is not only useful, but also relatively easy to carry out.

The first step of the MK transformation is illustrated for a two-dimensional square lattice of N sites in Figure 7.6. We begin by moving bonds to new locations, which will set us up to do a decimation transformation in the second step. There is no thermodynamic calculation involved here; we simply move bonds with the idea that as long as we retain all the bonds – just putting them in new locations – the restructured lattice should still have basically the same physics as the original. This is an uncontrolled approximation, and ultimately the validity of the method is determined by the results it gives. The basic idea can be applied to other lattices and other dimensions, but this example is perhaps the easiest to visualize.

On the left side of Figure 7.6 we show how the bonds are moved. Specifically, we move all the bonds in the interior of the 2 × 2 squares to the edges of the squares, as shown on the right side of the figure. Each vertical bond inside the 2 × 2 square is split in two, with one half moving to the right and one half moving to the left. Similarly, the horizontal bonds are split in two, with one half moving up and one half moving down. The net result is that all the bonds inside the 2 × 2 squares are moved away from their original locations, and the bonds on the edges of the 2 × 2 squares are strengthened to their new values $\widetilde{K}$. In this case it is clear, just by counting the bonds involved, that $\widetilde{K} = 2K$.

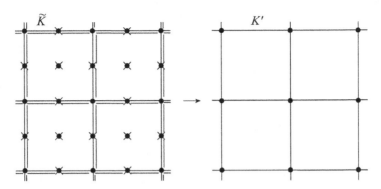

Figure 7.7 The second step of the MK transformation. The spins indicated with Xs on the restructured lattice are summed over in this step, leaving the renormalized lattice with a coupling K' and with $N' = N/4$ sites.

We are now ready to proceed with the second, nontrivial part of the MK transformation, which is illustrated in Figure 7.7. Notice that the 2×2 squares are perfectly set up for a decimation transformation – in fact, each side of a 2×2 square has the same basic unit as in Figure 7.2. Thus, we can sum over the middle spins on each side, indicated by the Xs, and generate a renormalized coupling, K', for the new lattice. The central spins in the 2×2 squares are disconnected from any other spins and can be summed over (as indicated by the Xs) to give a contribution of $\ln 2$ per disconnected spin to the free energy. Thus, by decimating all the spins indicated with Xs on the restructured lattice, we obtain the final lattice on the right in Figure 7.7, with $N' = N/4$ sites. The transformation can now be repeated as many times as desired.

Renormalized Coupling

As mentioned, we've already carried out the decimation transformation involved in step two of this transformation, and the results are given in Equation 7.1. It follows that the renormalized coupling for the MK transformation is

$$K' = \frac{1}{2}\ln[\cosh(2\widetilde{K})] = \frac{1}{2}\ln[\cosh(4K)]. \tag{7.7}$$

Notice that we've replaced $\widetilde{K}$ with $2K$, as is appropriate in this case. We present a plot of K' versus K in Figure 7.8.

There are significant qualitative differences between Figure 7.8 and the plot of K' for the 1-D chain in Figure 7.3. In the 1-D case, $K'(K)$ is always less than

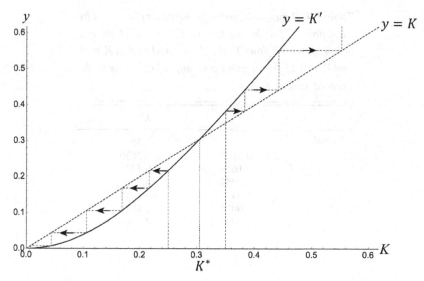

Figure 7.8 The renormalized coupling K' as a function of K for the MK trans-formation. The fixed point, where $K'(K^*) = K^* = 0.304\ldots$, represents the critical point for this system. Points with $K < K^*$ flow to the disordered fixed point at $K = 0$; points with $K > K^*$ flow to the ordered fixed point at $K \to \infty$.

K, and the flows of the RG transformation take all of the initial values of K to the attractive fixed point at $K = 0$. In Figure 7.8, on the other hand, K' starts out less than K for small values of K (large temperature), but becomes larger than K for large values of K (small temperatures). At an intermediate value of K there is a fixed point, K^*, where $K'(K^*) = K^*$. This is qualitatively different from what we saw in 1-D – in fact, this new behavior indicates a true phase transition (critical point) at $K'(K^*) = K^* = 0.304 \ldots$. For comparison, recall that the exact location of the critical point for a 2-D Ising model with nearest-neighbor couplings is $K_c = 0.4406 \ldots$, and the corresponding value found with mean-field theory is $K_c = 0.25$.

To see how we draw this conclusion – namely, that the fixed point at $K = K^*$ corresponds to a critical point – note first that the $K = 0$ attractive fixed point (often referred to as a "sink") represents a totally disordered state, since there is no coupling in the system in this case. An example of this kind of flow is shown in the middle column of Table 7.4, where we present iterations from the starting value $K = 0.25$. All values of $K < K^*$ flow toward $K = 0$, and hence are within the disordered phase of the system.

On the other hand, starting values of K that are greater than K^* all flow to infinity, which can be thought of as a second attractive fixed point, or sink. An

Table 7.4 *Renormalization-group flows generated by Equation 7.7 for the starting values* K = *0.25 (center column), which flows to the disordered sink at* K = *0, and* K = *0.35 (right-hand column), which flows to the ordered sink at* K → ∞.

n	$K^{(n)}$	$K^{(n)}$
0	0.25	0.35
1	0.2168	0.3829
2	0.1684	0.4421
3	0.1058	0.5520
4	0.0434	0.7635
5	0.0075	1.1816
6	0.0002	2.0166
.	.	.
.	.	.
.	.	.
∞	0	∞

example of this kind of flow is shown in the right-hand column of Table 7.4, where we present iterations from the starting value $K = 0.35$. In this limit, the coupling becomes infinitely strong, and hence all spins on the lattice point in the same direction – this is the totally ordered state. It follows that all values of $K > K^*$ are within the ordered phase of the system.

Between these two limits is the fixed point at $K = K^*$. This fixed point is referred to as a "nontrivial" or "relevant" fixed point because the value of K must be precisely equal to K^* to be at the fixed point; otherwise, the RG flows take you away to one of the sinks. Clearly, then, this fixed point separates the disordered phase from the ordered phase – in other words, it is the location of the phase transition between these two phases.

Going one step further with this identification, notice that the distance between nearest neighbors *increases* by a factor $b = 2$ with each iteration of the MK transformation, as can be seen on the right in Figure 7.7. This means, in turn, that the size of correlated patches of spins on the new lattice – measured in units of the new lattice spacing – *decreases* by a factor of b with each iteration. As the correlation length gets smaller, the system gets further away from criticality, as illustrated in Figure 5.5. At the critical point, however, the correlation length is infinite, and hence rescaling all lengths by a factor b has no effect. Therefore, the fixed point at K^* represents the system when it is critical and has infinite correlation length. We will return to the topic of the correlation length shortly, but first we present a simple and effective method to determine the numerical value of the fixed point of the transformation.

Reversing the Flow to Find the Fixed Point

As mentioned earlier, the relevant fixed point for the transformation of Equation 7.7 is at $K^* = 0.304$ .... How do we find this value? One way, of course, is to simply make a guess for K^* based on a plot like Figure 7.8, and then to refine the guess to get closer and closer to the fixed point. There's a systematic way to find the fixed point, however, referred to as *reversing the flow*, which we abbreviate with the subscript rf. This method generates flows that go automatically toward the desired fixed point rather than away from it; in effect, the rf transformation finds its own fixed point by simple iteration.

The basic idea for reversing the flow is illustrated in Figure 7.9. We know that $y = K'$ (dotted curve) generates flows that go away from K^*. Instead, suppose we produce a new transformation that *reflects* $y = K'$ about the line $y = K$. This reflected transformation is referred to as K_{rf} and is shown by the solid curve in Figure 7.9. Cobwebbing shows that K_{rf} does indeed produce reversed flows that go *toward* K^*.

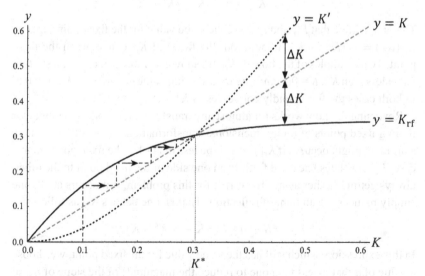

Figure 7.9 The MK transformation, $y = K'$ (dotted curve), is reflected about $y = K$ to produce the reversed flow transformation, $y = K_{rf}$ (solid curve). Cobwebbing shows that K_{rf} produces flows converging on the fixed point, K^*.

Table 7.5 *Renormalization-group flows for* K_{rf} *with two different starting values,* $K_{rf} = 0.1$ *and* $K_{rf} = 0.5$. *In both cases the flows converge to the fixed point at* $K^* = 0.304 ....$ *(The notation* $K_{rf}^{(n)}$ *indicates the nth iteration of* K_{rf}.)

n	$K_{rf}^{(n)}$	$K_{rf}^{(n)}$
0	0.1	0.5
1	0.1610	0.3374
2	0.2247	0.3140
3	0.2699	0.3075
4	0.2919	0.3056
5	0.3004	0.3049
6	0.3032	0.3047
.	.	.
.	.	.
.	.	.
∞	0.3046	0.3046

To carry out the reflection process mathematically, we note that if K' is *greater* than K by the amount $\Delta K = K' - K$, then we want K_{rf} to be *less* than K by the same amount, as indicated in Figure 7.9. Thus, we define K_{rf} as follows:

$$K_{rf} = K - \Delta K = K - (K' - K).$$

That's it. Notice that K_{rf} and K' have the same value for the fixed point; that is, $K_{rf}(K^*) = K'(K^*) = K^*$. In addition, the flows of K_{rf} converge on the fixed point. Two examples of the flow of K_{rf} are shown in Table 7.5, one for a starting value less than K^*, $K = 0.1$, and one for a starting value greater than K^*, $K = 0.5$. In both cases the flow rapidly converges to K^*.

Reversing the flow works for almost any transformation, and even works for finding fixed points of multidimensional transformations. The one case where a problem might occur is if K_{rf} has a slope less than -1 at the fixed point. In this case, K_{rf} oscillates back and forth from one side of the fixed point to the other, always getting farther away. To correct for this problem, if it occurs at all, one simply reduces the amount of reflection. That is, one defines K_{rf} as follows:

$$K_{rf} = K - a\Delta K = K - a(K' - K).$$

In this expression, which still has the same value for the fixed point, we choose a value of a that is less than one to reduce the magnitude of the slope of K_{rf} at the fixed point. A value of $a = 0.5$ or $a = 0.1$ generally works just fine in such cases.

Correlation Length Critical Exponent

As we've seen, the location of the relevant fixed point $K'(K^*) = K^*$ of the MK transformation gives the location of the system's critical point. It turns out that the slope of $K'(K)$ at the fixed point is also significant – in fact, it is related to the critical exponents of the system.

To see an example of this connection, let's start with the correlation length, ξ. This length diverges to infinity at the critical point, and in fact it diverges with an *exponential dependence* that is given by the following relation:

$$\xi(t) \sim t^{-\nu}. \tag{7.8}$$

In this expression, $t = |T - T_c|/T_c$ is the reduced temperature distance from the critical point, and $\nu > 0$ is the critical exponent.

Now, if we start a small distance away from the fixed point, at the location $K = K^* + \Delta K$, the renormalized coupling moves farther away by a factor of the derivative of K'. That is,

$$K'(K^* + \Delta K) = K^* + (\Delta K)\left(\frac{dK'}{dK}\right)_{K^*} + \dots.$$

Similarly, since t is connected to K by a simple inverse relationship, it follows that the value of t increases by the same factor as ΔK. That is,

$$t' \sim \left(\frac{dK'}{dK}\right)_{K^*} t. \tag{7.9}$$

Note that t plays a role similar to that of ΔK.

After one MK iteration, the correlation length is reduced by a factor b, and hence we can write

$$\xi(t') = b^{-1}\xi(t). \tag{7.10}$$

Combining Equations 7.8, 7.9, and 7.10 yields

$$\left[\left(\frac{dK'}{dK}\right)_{K^*} t\right]^{-\nu} = b^{-1}t^{-\nu}.$$

Rearranging, simplifying, and solving for ν, we find

$$\nu = \frac{\ln(b)}{\ln\left[\left(\frac{dK'}{dK}\right)_{K^*}\right]}. \tag{7.11}$$

Thus, the critical exponent ν is directly related to the derivative of K' with respect to K evaluated at the fixed point. In our case, we have $b = 2$,

$\left(\frac{dK'}{dK}\right)_{K^*} = 1.67\ldots$, and hence it follows that $v = 1.33\ldots$. This compares reasonably well with the exact result, $v = 1$.

Free Energy

As is usual with PSRG transformations, we can use K'_0 to calculate the free energy of the system. Returning to the expressions in Equation 7.1, we recall that the renormalized zero level is

$$K'_0 = \ln 2 + K'.$$

This is the same expression we had in 1-D, only in this case there is a different result for K' – namely, the one given in Equation 7.7. We now show how to use this expression to find the free energy.

First, we note that when we do the second step of the MK calculation, the decimation step, there are $N/4$ spins that are totally disconnected. It follows that they contribute $\frac{1}{4}\ln 2$ to the free energy per site. Next, the number of 2×2 squares is $N/4$, and each of them has one vertical and one horizontal side (to prevent overcounting). These sides each contribute K'_0 to the free energy per site. Finally, the renormalized lattice has $N/4$ sites and a coupling K'. With these results in mind, we can write the transformation of the free energy per site as follows:

$$\begin{aligned} f(K) &= \frac{1}{4}\ln 2 + \frac{1}{4} \cdot 2 \cdot K'_0 + \frac{1}{4}f(K') \\ &= \frac{1}{4}\ln 2 + \frac{1}{2}K'_0 + \frac{1}{4}f(K'). \end{aligned}$$

As a quick check, in the limit $K = 0$ we have $f(0) = \ln 2$ and $K'_0 = \ln 2$. Substituting these results in the preceding equation verifies that all the counting factors are indeed correct.

Carrying out additional steps of the transformation yields

$$f(K) = \frac{3}{4}(\ln 2)\left[1 + \frac{1}{4} + \frac{1}{4^2} + \ldots\right] + \frac{1}{2}\left(K' + \frac{1}{4}K'' + \frac{1}{4^2}K''' + \ldots\right).$$

Simplifying the summation, and expressing the result in terms of superscripts, we can write

$$f(K) = \ln 2 + \frac{1}{2}\sum_{n=1}^{\infty} \frac{K^{(n)}}{4^{(n-1)}}. \tag{7.12}$$

Table 7.6 *Applying the MK transformation to the initial value* K = *1.0. Notice that* K *flows to infinity, while the free energy converges to a finite value.*

n	$K^{(n)}$	$\ln 2 + \frac{1}{2} \sum_{i=1}^{n} \left(\frac{1}{4^{i-1}} \right) K^{(i)}$
0	1.0	0.6931
1	1.6536	1.5199
2	2.9606	1.8900
3	5.5747	2.0642
4	10.803	2.1486
5	21.259	2.1902
6	42.171	2.2107
.	.	.
.	.	.
.	.	.
∞	∞	2.2312

With this result we can calculate the free energy per site in the MK approximation for any starting value of K.

As an example, suppose we start with $K = 1.0$. In this case, $K' = 1.6536$. Thus, the first nontrivial approximation to the free energy is

$$\ln 2 + \frac{1}{2} K' = 1.5199.$$

The results from successive iterations are given in Table 7.6. Notice that the values of K increase with each iteration, as opposed to previous cases where K iterated toward $K = 0$. Of course, in this case our starting point is larger than K^*, and hence the flow is toward the sink at $K \to \infty$. Even though K is increasing without limit, the free energy converges quickly to a finite value of 2.2312 . . ., as compared to the exact value of 2.0003 . . . .

We present a plot of the MK free energy versus K in Figure 7.10, along with the exact free energy for comparison. Notice that both free energies start out at $\ln 2$ at $K = 0$, as expected for the totally disordered state. In addition, both results have a slope of $2K$ for large values of K, reflecting the fact that the system is totally ordered in that limit, and that there are two bonds of strength K per site for the infinite square lattice.

The MK free energy has a maximum error of about 20 percent, which occurs near the critical point. For large K there is a constant offset of about 0.213 . . . between the MK result and the exact free energy. Thus, as K increases, the percentage error decreases toward zero. The MK approximation is reasonably good in most respects but still leaves a bit to be desired. We address this next.

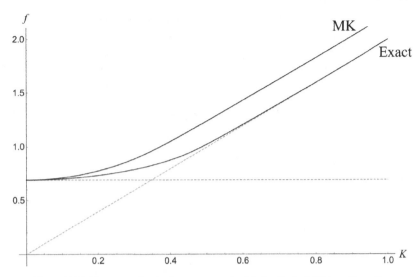

Figure 7.10 The curve for the free energy calculated with the MK transformation compared with that for the exact free energy for the square lattice.

Cluster-Decimation Approximation

In the MK approximation, the first step – the restructuring step – is where all the error occurs. After all, the second step – the decimation step – is exact. In addition, recall that the restructuring step is uncontrolled; there is no calculational basis for it, other than the idea that it probably makes sense to keep the total number of bonds on the lattice the same. What we desire, instead, is a way to restructure the lattice, while at the same time *preserving* the free energy per site – at least in an approximate way. We'll do this by considering finite clusters of spins as an approximation to the infinite lattice.

Consider, for example, a 2×2 cluster of spins on the original lattice. We can continue this cluster periodically over the lattice, as indicated in Figure 7.11. The free energy per site of this cluster is given by $f_{2 \times 2}(K)$, which we will calculate shortly. This is a reasonable approximation to the free energy of the original lattice.

Next, we again consider a 2×2 cluster, but this time on the restructured lattice. This cluster is also extended periodically, of course, with the result shown in Figure 7.11. This lattice has a coupling strength $\widetilde{K}$, and a free energy per site given by $\widetilde{f}_{2 \times 2}(\widetilde{K})$.

Now, the key step in this approach is to evaluate these free energies – each of which approximates the exact free energy on its lattice – and set them equal to one another. That is,

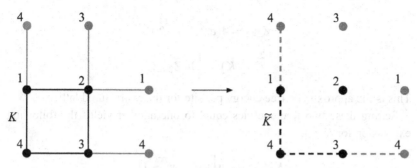

Figure 7.11 A 2 × 2 cluster extended periodically on both the original (K) and restructured $(\widetilde{K})$ lattices. The new coupling $\widetilde{K}$ ensures that the free energy per site of these two clusters is equal.

$$f_{2\times2}(K) = \widetilde{f}_{2\times2}(\widetilde{K}). \tag{7.13}$$

Solving this equation gives $\widetilde{K}$ as a function of K, replacing the simple MK prescription that $\widetilde{K} = 2K$. At this point the decimation transformation can be applied as before, only now using $\widetilde{K}(K)$ determined from Equation 7.13. Since this method combines the results of a cluster calculation and a decimation calculation, it is referred to as a cluster-decimation approximation (CDA).

Let's start by calculating the free energy per site of the 2 × 2 cluster on the original lattice. The reduced Hamiltonian for this system is

$$-\beta H = 2K(s_1 s_2 + s_2 s_3 + s_3 s_4 + s_4 s_1).$$

Notice that the periodic boundary conditions have doubled the strength of all the bonds. Summing over the configurations of all the spins yields the partition function and the free energy per site:

$$Z_{2\times2} = 2e^{8K} + 12 + 2e^{-8K}$$

$$f_{2\times2}(K) = \frac{1}{4}\ln Z_{2\times2}.$$

This is our approximate free energy per site for the original lattice.

Now, for the restructured lattice. The reduced Hamiltonian is

$$-\beta\widetilde{H} = 2\widetilde{K}(s_1 s_4 + s_4 s_3).$$

Again, the periodic boundary conditions have doubled the bond strength. The corresponding partition function and free energy per site are

$$\widetilde{Z}_{2\times2} = 4(e^{2\widetilde{K}} + e^{-2\widetilde{K}})^2$$

$$\widetilde{f}_{2\times2}(\widetilde{K}) = \frac{1}{4}\ln\widetilde{Z}_{2\times2}.$$

This is the approximate free energy per site for the restructured lattice.

Setting these two free energies equal to one another yields the following expression for $\widetilde{K}(K)$:

$$\widetilde{K}(K) = \frac{1}{4}\ln\left[y + \sqrt{y^2 - 1}\right].$$

In this expression, y is given by

$$y = \frac{1}{2}[1 + \cosh(8K)].$$

Combining these expressions gives the desired function $\widetilde{K}(K)$. A plot of the *bond strengthening factor*, defined as $\widetilde{K}(K)/K$, is given in Figure 7.12.

The first thing to notice about Figure 7.12 is that the MK prescription, $\widetilde{K}/K = 2$, always overstrengthens the bonds. This explains why the MK free energy is always greater than the exact free energy in Figure 7.10. In addition,

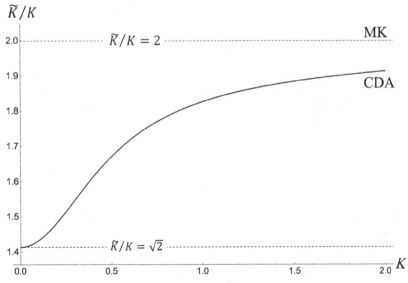

Figure 7.12 The bond-strengthening factor, $\widetilde{K}(K)/K$, for the cluster-decimation approximation (CDA). Notice that the upper limit for the CDA is the MK bond-strengthening factor (2), and the lower limit is the square root of this value ($\sqrt{2}$).

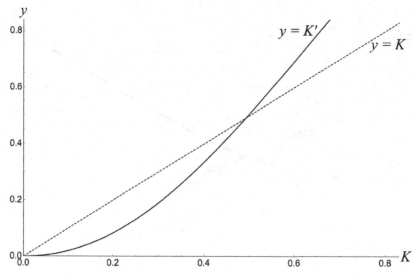

Figure 7.13 A plot of K' versus K for the CDA. Both the location of the fixed point and the slope of K' at the fixed point give improved results over the standard MK approach.

notice that the lower limit of the bond strengthening is the square root of the MK value, $\widetilde{K}/K = \sqrt{2}$. The CDA smoothly interpolates between these limits.

Using this result for $\widetilde{K}$ in the decimation step yields the renormalized coupling K' plotted in Figure 7.13. It has the same general shape as in the MK approximation result shown in Figure 7.8, but now the fixed point of the transformation is $K^* = 0.492$ .... This is considerably closer to the exact value of 0.440 ... than the MK result of 0.304 ....

As before, the slope of K' at the fixed point gives the critical exponent ν. In this case we find $\nu = 1.19$ .... This is to be compared with 1.33 ... for the MK approach and 1.0 for the exact result. Again, reasonably good improvement, and with a method that can be extended to larger clusters for better results.

We can also calculate the free energy, using the same method as for the MK approximation, but now using the improved result for $\widetilde{K}$. The free energy given by this calculation is shown in Figure 7.14, where it is compared with the exact free energy. Notice the very close agreement between the two, with only a slight gap between the results near the critical point. Clearly, the CDA calculation, with its free energy–preserving basis, does a much better job in this regard than the simple bond-moving prescription of the original MK approach, as shown in Figure 7.10.

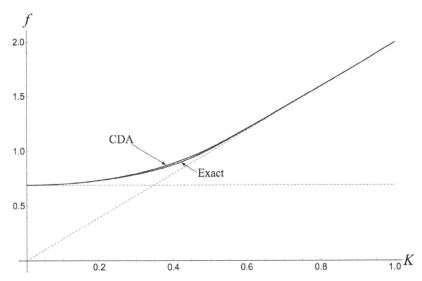

Figure 7.14 The reduced free energy per site from the CDA calculation compared with the exact free energy. The agreement is excellent, with only a slight difference near the critical point. The dashed lines show the small K limit of $\ln 2$ and the large K limit of $2K$.

Specific Heat

Finally, let's calculate the reduced specific heat per site, which is given by $c = C/Nk_B$. Recall from Equation 2.16 that this can be expressed as follows:

$$c = K^2 \frac{\partial^2 f}{\partial K^2}.$$

Perhaps the simplest way to calculate this quantity is to evaluate the second derivative numerically by calculating the free energy at a variety of values of K. Specifically, we can evaluate the following expression to calculate c at the location K:

$$c = K^2 \frac{f(K + \Delta K) - 2f(K) + f(K - \Delta K)}{\Delta K^2}.$$

Using $\Delta K = 0.001$ gives values that are perfectly acceptable for plotting.

Results from such a calculation are presented in Figure 7.15 and compared with the exact specific heat from the Onsager solution. Again, we see significant improvement of the CDA method as compared with the MK transformation, which is also shown in Figure 7.15. Not only is the location of the critical point improved, but the shape and maximum value of the specific heat are much better.

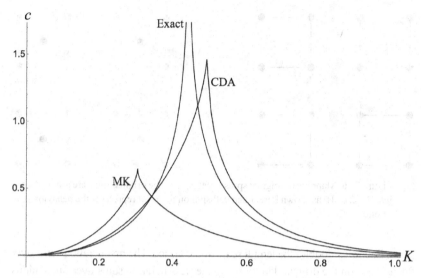

Figure 7.15 Reduced specific heats per site for the square lattice, comparing the Exact, CDA, and MK results.

To summarize, the MK approach is fairly simple to implement, and it gives reasonably good results. The CDA improves on the standard MK by putting the restructuring step on a calculational basis that is still rather simple to use. It also allows a path for systematic improvement by using larger clusters.

7.3 The Cell-Cluster Renormalization-Group Method

Another way to implement a PSRG transformation is referred to as the *cell-cluster method*. This kind of transformation doesn't involve moving bonds or performing decimations, as in the other methods in this chapter. Instead, the basic idea is to collect groups of spins on the original lattice into "cells," like the 2×2 squares shown in the left side of Figure 7.16. The transformation then replaces these cells with "cell spins" on a new lattice with a larger length scale, as illustrated on the right side of Figure 7.16. Preserving the partition function generates renormalized couplings between the cell spins and sets a new zero level for the free energy, as before.

The first step in the procedure indicated in Figure 7.16 is to implement a *mapping* between the original spins in a 2×2 cell and a cell spin that represents the overall behavior of the cell. The next iteration of the transformation uses the same procedure to map 2×2 groups of *cell spins* onto new sets of "super cell"

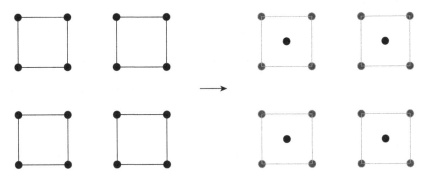

Figure 7.16 Mapping of original spins to cell spins. The original spins are grouped into 2 × 2 cells, as shown here. Each cell spin on the right represents the behavior of one 2 × 2 cell.

spins. Each step doubles the length scale of the lattice. The first iteration sums over fluctuations on the original length scale, the next iteration sums over fluctuations on twice that length scale, the next iteration includes fluctuations on four times that length scale, and so on. In this way, the RG transformation eventually sums fluctuations on all length scales. At the critical point, where the correlation length is infinite, the transformation leaves the system unchanged – that is, it produces a fixed point.

Projection Operator

To do the mapping of original spins to cell spins we use what is referred to as a *projection operator* to connect the spins. The simplest choice for the projection operator is *majority rule*, in which the cell spin represents the majority of the original spins. If the original spins are split equally between plus and minus, the mapping assigns equal weight to plus and minus cell spins. This projection is presented in Table 7.7.

We can write a mathematical expression for the projection operator as follows:

$$P(\{s_1, s_2, s_3, s_4\}, s') = \frac{1}{2}(1 + \text{sign}[s_1 + s_2 + s_3 + s_4]s').$$

In this equation, $\{s_1, s_2, s_3, s_4\}$ represents a configuration of the four spins in one of the 2 × 2 cells, and s' represents the value of the cell spin. The *sign* function returns $+1$ for a positive value of the argument, -1 for a negative value, and 0 when the argument is equal to zero. It is easily verified that P gives the values presented in Table 7.7.

Table 7.7 *The majority-rule projection. This projection maps configurations of original spins, $\{s_i\}$, to a cell spin, s'. If the majority of the original spins are $+$ or $-$, the mapping is to a $+$ or $-$ cell spin, respectively. If the original spins are half $+$ and half $-$, the projection goes equally to a cell spin that is $+$ or $-$.*

s_i	$s' = +1$	$s' = -1$
$++$ $++$	1	0
$++$ $+-$	1	0
$++-+$ $--,+-$	½	½
$+-$ $--$	0	1
$--$ $--$	0	1

Preserving the Free Energy

Now, the next step in constructing this transformation is to *preserve the partition function* as we go from the original lattice to the cell lattice. To do this, we must consider only a finite *cluster* of cells, for which we can carry out an explicit calculation. The simplest nontrivial cluster consists of two 2×2 cells, as indicated in Figure 7.17. Notice that the cells are continued periodically across the original lattice, as shown by the numbering system on the left. Similarly, the two cell spins are continued periodically on the cell lattice, as shown on the right.

To preserve the partition function, we start with the Hamiltonian for the two-cell cluster:

$$-\beta H = K[s_1(s_8 + s_2) + s_2(s_7 + s_1) + s_3(s_2 + s_4) + s_4(s_1 + s_3)$$
$$+ s_5(s_4 + s_6) + s_6(s_3 + s_5) + s_7(s_6 + s_8) + s_8(s_5 + s_7)].$$

Next, we write the Hamiltonian for the two cell spins:

$$-\beta H' = K'[s_1'(s_2' + s_1') + s_2'(s_1' + s_2')] + 2K_0'.$$

Notice that K_0' is the zero-level contribution per site.

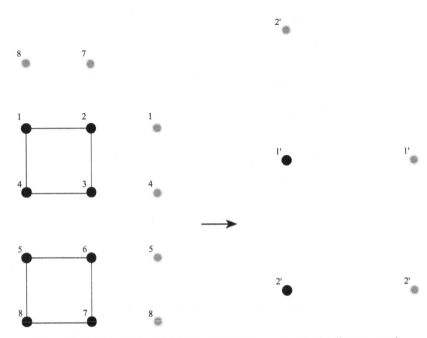

Figure 7.17 The cell-cluster transformation. In this case, two 2×2 cells are mapped to two cell spins. Both the original and cell lattices are continued periodically.

Now, if we set $s'_1 = +1$ and $s'_2 = +1$, we find

$$4K' + 2K'_0 = \ln[Z_{++}].$$

The partial partition function Z_{++} is given by the following:

$$Z_{++} = \sum_{\{s_1, s_2, s_3, s_4, s_5, s_6, s_7, s_8\}} P(\{s_1, s_2, s_3, s_4\}, +1) P(\{s_5, s_6, s_7, s_8\}, +1) e^{-\beta H}$$

$$= e^{16K} + \frac{21}{2} e^{8K} + 10 e^{4K} + 25 + 10 e^{-4K} + 7 e^{-8K} + \frac{1}{2} e^{-16K}.$$

Similarly, if we set $s'_1 = +1$ and $s'_2 = -1$, we find

$$2K'_0 = \ln[Z_{+-}].$$

The partial partition function in this case is

$$Z_{+-} = \sum_{\{s_1, s_2, s_3, s_4, s_5, s_6, s_7, s_8\}} P(\{s_1, s_2, s_3, s_4\}, +1) P(\{s_5, s_6, s_7, s_8\}, -1) e^{-\beta H}$$

$$= \frac{7}{2} e^{8K} + 14 e^{4K} + 25 + 14 e^{-4K} + 7 e^{-8K} + \frac{1}{2} e^{-16K}.$$

Thus, the renormalized Hamiltonian is characterized by the following:

$$K' = \frac{1}{4} \ln \left[\frac{Z_{++}}{Z_{+-}} \right]$$

$$K_0' = \frac{1}{2} \ln[Z_{+-}].$$

We now know the new coupling and the new zero level.

The transformation is completed by writing out the free energy connection between the lattices. Noting that the cell lattice has ¼ the number of sites of the original lattice, we have

$$f(K) = \frac{1}{4} K_0'(K) + \frac{1}{4} f(K').$$

Extending this result to multiple iterations, we can write

$$f(K) = \frac{1}{4} K_0'(K) + \frac{1}{4^2} K_0'(K') + \frac{1}{4^3} K_0'(K'') + \ldots = \sum_{n=0}^{\infty} \frac{1}{4^{(n+1)}} K_0' \left(K^{(n)} \right).$$

Our cell-cluster transformation is fully defined now, and we can calculate the renormalized couplings, the free energy, and all other thermodynamic quantities of interest.

Cell-Cluster Results

The fixed point of the renormalized coupling, K', gives the location of the critical point. We find

$$K'(K^*) = K^* = 0.391 \ldots$$

This is a good approximation to the exact value of $0.440 \ldots$. The derivative of K' at the fixed point yields the critical exponent v. In this case,

$$v = \frac{\ln(2)}{\ln \left[\left(\frac{dK'}{dK} \right)_{K^*} \right]} = 1.13 \ldots$$

Again, quite good agreement with the exact value of 1.0.

We can also calculate the reduced free energy per site as a function of K. The results are given in Figure 7.18. There is only a very slight difference near the critical point between the approximate and exact free energies. These results are similar to those obtained with the CDA calculations in the previous section, and both are much improved over the MK results.

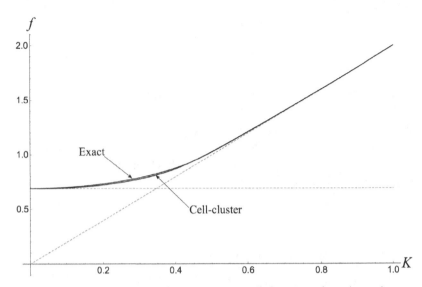

Figure 7.18 Comparing the free energy of the cell-cluster transformation to the exact free energy for the 2-D square lattice. The agreement is excellent, with only slight differences near the critical point.

A more stringent test of the transformation is provided by the specific heat. In Figure 7.19 we present the cell-cluster and exact specific heats for comparison. The RG result tops out at a finite maximum value but is still in quite good agreement, both qualitatively and quantitatively, with the exact specific heat. In contrast, the MF results, also shown in Figure 7.19, give results that are qualitatively different – namely, identically equal to zero for small K. Clearly, the RG approach has succeeded to a much greater degree than the mean-field calculations.

We've seen that PSRG techniques provide an improved approach to approximate calculations of Ising systems, as compared with mean-field calculations. In addition, the PSRG methods can be extended to give improved results, and can be applied to lattices with different structures, like hexagonal or triangular, as well as lattices in different dimensions.

7.4 Problems

7.1 Consider the $b = 3$ decimation transformation for the 1-D chain lattice shown in Figure 7.20. The lattice has nearest-neighbor interactions of strength K. **(a)** Derive the transformation equations for K' and K'_0 in

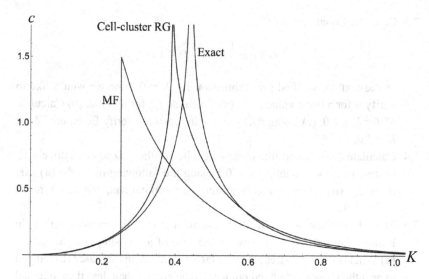

Figure 7.19 Comparisons of the reduced specific heat per site. The cell-cluster and exact results are quite similar. The mean-field (MF) specific heat is identically zero for K smaller than the critical value, which is qualitatively different from the exact specific heat.

terms of the partial partition functions Z_{++} and Z_{+-}. **(b)** Derive expressions for Z_{++} and Z_{+-}. **(c)** What is the transformation equation relating the free energy of the original lattice, $f(K)$, to the free energy of the primed lattice, $f(K')$?

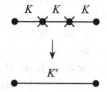

Figure 7.20 Problems 7.1 and 7.2.

7.2 Consider the $b = 3$ 1-D transformation in Problem 7.1. **(a)** For a given finite value of K, do you expect K' for $b = 3$ to be greater than, less than, or equal to K' for $b = 2$? Explain. **(b)** For a given value of K, do you expect K'_0 for $b = 3$ to be greater than, less than, or equal to K'_0 for $b = 2$? Explain. **(c)** Check your results in parts (a) and (b) with explicit calculations for $K = 1.0$.

7.3 Consider Equation 7.4:

$$f(K) = \frac{1}{2}K_0' + \frac{1}{2}f(K').$$

In the text we verified this expression for $K = 0$; now we would like to verify it for a finite value of K. **(a)** Calculate K_0' for $K = 2.0$. **(b)** Calculate K' for $K = 2.0$. **(c)** Using $f(K) = \ln 2 + \ln(\cosh K)$, verify Equation 7.4 for $K = 2.0$.

7.4 Calculate the reduced free energy per site for the 2-D square lattice with nearest-neighbor coupling $K = 0.5$ using the following methods: **(a)** MK ($b = 2$), **(b)** CDA, **(c)** cell-cluster. For comparison, the exact result is $f = 1.0257\ldots$.

7.5 **MK Transformation in 3-D** Consider the MK transformation in Figures 7.6 and 7.7, only now for the case of a 3-D simple cubic lattice. **(a)** What is the bond strengthening factor $\widetilde{K}(K)/K$ in this case? **(b)** Do you expect the value of the fixed point K^* to be greater than, less than, or equal to the value found for the 2D lattice? Explain. **(c)** Use the reverse flow transformation to find K^*.

7.6 **Tyrant's Rule** In this problem, we consider a different type of projection from site spins to cells spins referred to as *Tyrant's Rule*. In this case, the value of the cell spin is simply equal to one of the site spins – the tyrant. As a specific example, let's say that the site spin in the upper left corner of the cell is the tyrant. **(a)** Give a table like Table 7.7 that defines this projection. **(b)** Find Z_{++} and Z_{+-} for this transformation. **(c)** Find the fixed point K^* for this transformation.

7.7 **Egalitarian Rule** In this problem, we consider a different type of projection from site spins to cells spins referred to as *The Egalitarian Rule*. In this case, each site spin contributes equally to the cell spin. For example, if four site spins are $+$, the system maps to a $+$ cell spin with weight 1 and to a $-$ cell spin with weight 0. On the other hand, if three site spins are $+$ and one is $-$, the system maps to a $+$ cell spin with weight $\frac{3}{4}$, and to a $-$ cell spin with weight $\frac{1}{4}$. **(a)** Give a table like Table 7.7 that defines this projection. **(b)** Find Z_{++} and Z_{+-} for this transformation. **(c)** Find the fixed point K^* for this transformation.

Bibliography

Goldstein, R. E. and Walker, J. S. Thermodynamic functions and critical properties from a cluster-decimation approximation. *J. Phys. A*, **18**, 1275, 1985.

A presentation of the cluster-decimation approximation (CDA) and comparisons of it with the standard Migdal–Kadanoff approach.

Ising, E. Beitrag zur Theorie des Ferromagnetismus. *Z. Phys.*, **31**, 253, 1925.

The original paper on the "Ising" model. It presents the results of Ising's doctoral thesis and suggests that the model may be too simple to show a phase transition, even in higher dimensions – a conclusion that, thankfully, proved to be overly pessimistic.

Kadanoff, L. P. Notes on Migdal's recursion formulas. *Ann. Phys.*, **100**, 359, 1976.

This paper sets out the basic features of the Migdal–Kadanoff approximation. Migdal's original paper, which was rather obscure and published in a Russian journal, didn't receive much attention until Kadanoff gave it this more accessible presentation.

Kramers, H. A. and Wannier, G. H. Statistics of the two-dimensional ferromagnet. Part I. *Phys. Rev.*, **60**, 252, 1941.

Kramers, H. A. and Wannier, G. H. Statistics of the two-dimensional ferromagnet. Part II. *Phys. Rev.*, **60**, 263, 1941.

These papers set out the results of duality, which gives the exact location of the critical point for the two-dimensional, nearest-neighbor Ising model on a square lattice.

Landau, L. D. and Lifshitz, E. M. *Statistical Physics*. Elsevier, 3rd edition, 1980.
A standard text on statistical mechanics.

Onsager, L. Crystal statistics. I. A two-dimensional model with an order-disorder transition. *Phys. Rev.*, **65**, 117, 1944.

Onsager's famous paper detailing his solution to the two-dimensional Ising model for the case of a square lattice with nearest-neighbor interactions. This is one of the most cited papers in all of theoretical physics, though only few have read it, and fewer still have worked through the calculations. Ken Wilson remarked that after finishing his doctoral thesis in high-energy physics he set himself the task of learning the Onsager solution – an exercise that led him to develop the Nobel Prize–winning renormalization group in statistical mechanics.

Pathria, R. K. *Statistical Mechanics*. Butterworth, Heinemann, 2nd edition, 1996.

A good introduction to many topics in statistical mechanics, including mean-field theory and Landau theory.

Peierls, R. On Ising's model of ferromagnetism. *Proc. Cambridge Phil. Soc.*, **32**, 477, 1936.

In this paper, Rudolf Peierls presents his rigorous argument showing that the two-dimensional Ising model with nearest-neighbor interactions does indeed have a phase transition – which opened the door to further studies of the model. I was fortunate enough to know "Rudy" when I was a graduate student, and only years later learned that he was actually Sir Rudolf Peierls, so named in honor of his nuclear physics research during World War II. His life story is told in a fascinating autobiography, *Bird of Passage: Recollections of a Physicist*, Princeton University Press, 1985.

Schick, M., Walker, J. S., and Wortis, M. Phase diagram of the triangular Ising model: Renormalization-group calculation with application to adsorbed monolayers. *Phys. Rev. B*, **16**, 1977.

An early cell-cluster, renormalization-group paper showing the utility of the method for applications in pure theory, as well as in making comparisons with experiments.

Stanley, H. E. *Introduction to Phase Transitions and Critical Phenomena*. Oxford University Press, 1971.

A classic text covering the state of the art in phase transitions and critical phenomena before the introduction of the renormalization group.

Walker, J. S. and Vause, C. A. Lattice theory of binary fluid mixtures: Phase diagrams with upper and lower critical solution points from a renormalization-group calculation. *J. Chem. Phys.*, **79**, 2660, 1983.

An application of a modified Migdal–Kadanoff approximation to binary liquid mixtures with reentrant phase transitions.

Wilson, K. G. Renormalization group and critical phenomena. I. Renormalization group and the Kadanoff scaling picture. *Phys. Rev. B*, **4**, 3174, 1971.

An early paper by Kenneth Wilson outlining his development of the renormalization-group technique in statistical mechanics.

Yang, C. N. The spontaneous magnetization of a two-dimensional Ising model. *Phys. Rev.* **85**, 808, 1952.

This paper derives the exact formula for the zero-field magnetization of the two-dimensional Ising model.

Index

Printed in the United States
by Baker & Taylor Publisher Services

Printed in the United States
by Baker & Taylor Publisher Services